BIOREGIONAL SOLUTIONS

D0512912

Pooran Desai and **Sue Riddlestone** are co-founders and Directors of BioRegional Development Group. Pooran studied physiological sciences at Oxford University. He founded the BioRegional Charcoal Company and has worked to revive south London's lavender industry. With architect Bill Dunster he initiated the BedZED eco-village. Pooran is a member of the UK government's Advisory Committee on Consumer Products and the Environment, chairing its homes and environment group. Sue, a mother of three, formerly practised as a nurse. She led a team to develop the MiniMill, new small-scale technology to produce paper pulp, and is Director of BioRegional MiniMills Ltd. She initiated Local Paper for London and research into UK hemp textiles. Sue is a member of the London Sustainable Development Commission.

Schumacher Briefing No. 8

BIOREGIONAL SOLUTIONS

For Living on One Planet

Pooran Desai and Sue Riddlestone

published by Green Books
for The Schumacher Society

First published in 2002
by Green Books Ltd
Foxhole, Dartington, Totnes,
Devon TQ9 6EB
www.greenbooks.co.uk
greenbooks@gn.apc.org

for The Schumacher Society
The CREATE Centre, Smeaton Road,
Bristol BS1 6XN
www.oneworld.org/schumachersoc
schumacher@gn.apc.org

in association with
The BioRegional Development Group
BedZED Centre, 24 Helios Road
Wallington, Surrey SM6 7BZ
www.bioregional.com

Figures 6 & 7 and photography by Katrina Stewart
Cover design by Rick Lawrence

Printed by J.W. Arrowsmith Ltd, Bristol, UK

Main text paper: 20% Essex hemp/80% London recycled paper
made in Hemel Hempstead; colour plates printed on 100%
recycled paper; cover board made from 75% recycled fibre.

A catalogue record for this publication
is available from the British Library

ISBN 1 903998 07 7

Contents

Acknowledgements

We are grateful to the many individuals and organisations who have worked with us to develop and implement the BioRegional solutions described in this Briefing. Thanks are due to all the dedicated people who have worked at BioRegional and its associated companies, especially those still with us from the early days of BioRegional, including Sarah Alsen, Nicole Lazarus and Nicola Davies; our patron Professor Sir Ghillean Prance; our Board of Trustees; our Honorary Trustee Jean-Paul Jeanrenaud; Geoffrey Cox (BioRegional Charcoal Ltd) and Alan Knight (B&Q) who were instrumental in the charcoal initiative; Bill Dunster and his team at Bill Dunster Architects, Chris Twinn (Arup), Jonathan Deans, Mathew Mackey, Chris Welch, Eric Muscat and Andy Cooper (Gardiner and Theobald) and Lachlan McDonald (Ellis and Moore) for the design and construction of BedZED; Dickon Robinson and Malcolm Kirk of The Peabody Trust without whom BedZED would have remained on the drawing board; Rev. Andrew Roland and June and Peter Thomas; Chas Ball (Smart Moves). The DTI, Clive Ward, John Leyland, Trevor Dean, Stefan Kay, David Wei and the UK papermakers for their work on MiniMills; M-Real, London Recycling, Direct Line Insurance, George Alagiah and Ken Livingstone and all those involved in the Local Paper for London initiative, including Tony Hart for his work on the life cycle assessment; Bob Franck, Chris Older, Hemcore, Harry Gilbertson, Kate Fletcher, Angela Shipp and Katharine Hamnett who worked with us to demonstrate the potential for UK hemp textiles; HM Prison Downview, Yardley of London and the volunteers who helped establish the lavender fields; to those at the Croydon urban forestry site including Simon Levy, Bill Bailey and Bob Roseberry; Bernadette Vallely and the Women's Environmental Network; Tom Brake MP; Professor David Hughes; Mike Kirby; Professor Roland Clift; Peter Jones; Steve Parry; Vera Elliot; Alison Lucas and Paul King; and to David, Tom and Sarah for their support.

We are also pleased to have the opportunity to thank the many charitable trusts and funding bodies who have supported us in the realisation of our ideas, including Biffaward, Body Shop Foundation, Bridge

House Estates Trust Fund, Cecil Pilkington Charitable Trust, Charities Aid Foundation, Charlotte Bonham-Carter Charitable Trust, Countryside Agency, Cuthbert Horn Trust, Drapers' Charitable Trust, EB Nationwide Limited (Shanks First), Environmental Action Fund, Ernest Cook Trust, Esmée Fairbairn Foundation, Forestry Commission, individuals through the Give As You Earn scheme, Girdlers Company, Heritage Lottery Fund, Housing Corporation, J. Paul Getty Jr. Charitable Trust, J.J. Charitable Trust, Konrad Zweig Charitable Trust, London Borough of Sutton, London Cycling Campaign, Lyndhurst Settlement, Manifold Trust, Mark Leonard Trust, Ministry of Agriculture, Fisheries and Food, Mitchell Trust, Mrs C. Eynon No. 8 Settlement, National Grid Community 21 Awards, Naturesave Trust, Norlands Foundation, Partners in Innovation Programme, Polden Puckham Foundation, Prince of Wales Trust, Robert Kiln Charitable Trust, Rufford Foundation, Schumacher Society, SEED Fund, SEI Trust, Shell Better Britain Campaign, Transco Grassroots Award Scheme, Tudor Trust, World Wide Fund for Nature International and the World Wide Fund for Nature UK.

Finally, thanks to WWF-UK for part funding the production of this briefing and to Herbert Girardet for encouraging us to write it.

Achieving sustainability is a massive challenge for humanity, requiring major changes in the way we think and act. Neither of those things is impossible, but I often wonder whether anything short of a series of human-induced catastrophes will be sufficient to jolt us out of our complacency. We have simply come to take all the services provided by the natural world for granted, refusing to accept that natural systems have limits and that all of our actions have consequences.

It is not that the issues are somehow unknown. One of the successes of the environmental movement, over the last thirty years or so, has been to generate widespread awareness of the problems we face. Nor is there any shortage of sustainability theory. The concept of living within the long-term capacity of our planet to support us receives widespread recognition and a fair degree of acceptance. We are just not very good at putting the theory into practice in any joined up, coherent, systematic and consistent fashion.

The capacity to produce innovative, mainstream solutions, and then make them work in the real world, is much the weakest link in our approach to sustainability. In these circumstances, organisations like BioRegional have a vital role to play in dispelling short-sightedness, challenging the conventional and the cautious and providing inspiration.

BioRegional's common sense approach of using local resources from farming, forestry, solar energy and recycled waste to meet more of our everyday needs brings benefits to society in numerous ways. The transport of goods is reduced, cutting fossil fuel consumption, reducing our contribution to global warming and improving air quality for a positive effect on health. The creation of local employment leads to more diverse and healthy local economies. And the physical proximity of producer and consumer can help protect local communities from some of the negative consequences of the global economy and lead to increased accountability.

BioRegional's charcoal project is a particularly good example of sustainability in action and demonstrates that the starting point for doing things well is to stop doing them badly. It makes no sense at all for the UK to be importing 60,000 tons of charcoal every year while this country has ample coppice woodland which can be brought back to life in the course of making excellent charcoal. A significant proportion of our imported charcoal emanates from tropical and mangrove forests in the developing world, and is then transported long distances to our barbeques. Yet these destructive effects can be substituted directly by a real benefit to Britain's woodland wildlife as birds, flowers and butterflies flourish in woodland that is coppiced regularly, sometimes for the first time in generations. Local production, with local employment, for local needs certainly makes sense in this case and, I suspect, in many others.

BioRegional has also developed many other practical solutions to problems of sustainability, in fields as diverse as paper manufacture and urban sustainability. I hope that this book will help to bring all those projects to a wider audience, providing them with inspiring proof of what can be achieved with imagination and determination.

In a world dominated by the short-term, the need for constructive thinking about our long-term future on this planet, based on wisdom and enduring values, has never been greater. Organisations like BioRegional, with the intellectual capacity and practical resolve to keep on coming up with solutions, have a hugely important role to play, and deserve all our support. I certainly wish them every possible success.

Preface

For the last eight years I have seen the BioRegional Development Group define and develop their ground-breaking ideas and implement more and more of their remarkable projects. There's been much *talk* about local sustainable development, but here is inspired and effective *action* of a most significant kind. BioRegional have taken a leading role in showing the way to a sustainable future and we are delighted to publish this book about their work.

The concept of bioregionalism has received much attention all over the world in recent years. The American author Kirkpatrick Sale describes it with these words: ". . . Imagine a society divided into territories and communities where love of place is an inevitable by-product of a life mindful of natural systems and patterns experienced daily— however far removed this may seem just now from the gigantic, destructive society around us." (*The Ecologist*, 22 February 2002)

In their own work, BioRegional have taken a slightly different approach. They have, first and foremost, emphasised the importance of utilising local resources for sustainable living and working, implementing the well-known concept of thinking globally and acting locally in practical terms. It is particularly to their credit to point out that even in an urban environment a great variety of resources can be found and sustainably utilised.

In a globalising world it is of great significance to show how local biological resources can be a viable basis for sustainable livelihoods. Implementing BioRegional's concept of the *network economy*, which brings together small local producers with effective marketing, will have tremendous repercussions in the coming years. An important point about publishing this book is to indicate that these methodologies can be replicated by other local groups elsewhere.

BioRegional's work very much reflects the ideas and approaches set out in E.F. Schumacher's book *Small is Beautiful*. This seminal text has sold 4 million copies and has been deeply influential worldwide; next year (2003) will be the 30th anniversary of its publication. Schumacher's ideas on appropriate technology have been widely adopted in developing countries, but less so in Europe and America. BioRegional's work has made up for some of that deficiency; developing small-scale technologies for papermaking, charcoal production and energy supply to buildings is of great significance for a sustainable future.

Bioregional development is not only about earnest attempts to reduce our ecological footprints and to fit in with the earth's long-term carrying capacity, but also about creating better relationships between people. The distances many of us have to travel every day are also tearing apart social relationships and isolating us. Proximity helps us to be more active in creating new communities and friendships, an issue that is most important for children.

The Beddington eco-village in South London, completed in July 2002, is a uniquely important statement on how housing and workspaces with modern amenities can be created which have minimal environmental impacts. Already BedZED is seen as a benchmark of sustainable urban development. I only hope that the very success of the project will not mean that the flood of visitors will be difficult to bear for the people who have chosen to live there.

BioRegional's work show just how much is possible under the auspices of the current economic and political system. To define what is possible under the status quo and what requires systems change in order to create a sustainable and equitable world will be the subject of another Schumacher Briefing in due course.

BioRegional have enjoyed the support of a great many organisations, and this is a significant indication of the importance with which their work is regarded. The same, of course, applies to the fact that the Prince of Wales has written the Foreword to this Briefing, for which we are very grateful indeed. Once again he has shown his extraordinary capacity to identify important new initiatives.

The Schumacher Society is proud to be associated with the pioneering work of the BioRegional Development Group. I am certain that in the coming years we will hear a lot more about them, and we hope that this Briefing will help to spread the word.

Herbert Girardet
Chairman, Schumacher Society
Commissioning Editor, Schumacher Briefings

Introduction

Can we build high quality eco-homes and sell them for the same price as conventional ones? Can we supply local wood products to a multi-national retailer whilst increasing the biodiversity of woodlands at the same time? Can we develop technology for small-scale, regional pulp and paper production, and produce paper at a competitive price? Can we use abandoned urban land to revive a traditional industry and sup-ply an international company? Can local authorities manage street trees as a productive forest? Working with a wide range of partners, these are the sorts of questions for which BioRegional Development Group have been developing solutions.

Ecological footprinting tells us that to be environmentally sustain-able, the UK needs to reduce its consumption of fossil fuels and virgin materials by two-thirds. To achieve a two-thirds reduction in consump-tion will require us to develop sustainable ways of living which are accessible and attractive to the mainstream of our society. We will need to provide sustainable products and services which can compete within a market which does not disadvantage them. We will need to be cre-ative and courageous.

The BioRegional approach is a practical expression of thinking glob-ally and acting locally. It is about bringing local sustainability into the mainstream. Its key principles are:

- Using local renewable and waste resources to meet more of our everyday needs

- Delivering sustainability through the market for products and services

- Living within our planetary means and leaving some space for wildlife and wilderness.

This involves the application of efficient small-scale technology and sys-tems to allow local products and services to enter the mainstream of our economy.

Localising the supply of products and services enables us to increase local recycling and reduce unnecessary transport. It also allows us to create healthy and more stable regional economies, protected from the destructive swings of globalisation. It promotes equitable development between nations. There is of course a place for international trade, but we may need to reconsider indiscriminate global competition where it damages the environment, particularly where it contributes to global warming. We believe our FEET index ('Foreign Exchange Earned per Tonne of Transport CO_2') can help define policies to minimise the environmental damage of long-distance trade.

Where we have needed new technologies to implement our projects we have teamed up with companies supplying them, or found partners in industry, government and research institutions to develop solutions. Examples include the small-scale waste wood-fired power plant we are installing at our BedZED eco-village, our MiniMill technology for pulping straw for paper, new more efficient charcoal-making kilns and new technology for extracting textile quality fibre from hemp.

We found that beyond new technology we also need new production and delivery systems. We have pioneered the concept of 'network production'—for instance, forming BioRegional networks of producers of wood products to supply a national retail chain. These networks marry the environmental and social benefits of local supply with the financial and consumer benefits of quality control and centralised marketing.

In our work we have found that the major barriers to sustainable development are institutional, arising from a compartmentalised system of decision-making in both public and private sectors—a failure to be able to think and act in a joined up way. One department in a company or government often promotes policies which conflict with those of another department. For instance, the corporate affairs department may see the great advantages of stocking local products, at the same time as the buying department signs global purchasing contracts. Or, one government department introduces carbon taxes on fossil fuels, when another gives planning permission for new airport terminals.

Better frameworks for decision-making are needed. Emerging disciplines like ecological footprinting quantify the global resources available to humanity on a renewable basis. This methodology indicates that since the 1970s we have been consuming more than our planet can support and that we need to reinvent our relationship to the earth. Sustainability needs to be based on common sense and good science, but it is by no means rocket science. New and old solutions are everywhere—often staring us in the face. Together we need to find ways of creating a decent standard of living for everyone, while also leaving some space for wilderness and a rich diversity of wild plant and animal life.

As we complete this book we are moving our office and homes into BedZED with the excitement and optimism of pioneers, and a sense that with our colleagues and partners we are colonising the future. We are of course building on 30 years of work by environmentalists from all walks of life to get sustainability accepted as a concept. We hope that our work is helping to take the environmental movement into a new phase.

Pooran Desai and Sue Riddlestone
12 May 2002

Chapter 1
Joined Up Locally

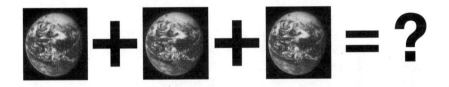

If everyone on the planet consumed as much as the average person in the UK, we'd need three planets to support us.[1] Our target in the UK must therefore be to reduce our consumption of raw materials and fossil fuels—our ecological footprint—by two-thirds. Only then can we say we are living sustainably and giving people around the world an equal chance to share in the earth's resources. Meeting more of our everyday needs from local renewable and reclaimed sources—bioregional development—can help us to achieve this target. Although this may seem unattainable at first, in our own experience from our own projects we know it is possible and can be achieved in the mainstream. At the same time we can increase our quality of life, re-create communities, promote accountability and leave space for wildlife and wilderness.

We live in a market economy and therefore the approach we have taken has been based on providing sustainable products and services. As we transform into a sustainable society, companies will need to supply products and services that allow us all to live within our means on this planet. They will need to be designed in such a way that they don't destroy natural systems. They will also need to promote social equity. We will not be able to live sustainably on a small planet with great divisions between rich and poor, causing resentment and unrest, where the rich are destroying the planet through overconsumption and the poor through desperation.

The approach we have taken arises out of the philosophy of bioregionalism[2] in which natural factors, such as local ecology and climate, rather than politics and trade, determine the way people live.* At BioRegional Development Group, however, we have been keen to engage with the market on its own terms and to link ourselves back to the local environment and the earth's natural nutrient and energy cycles. We value technology and the marketplace, but recognise that these can only bring long term benefits when they are linked to natural cycles—i.e. when we work with, rather than against, nature. If not, we will continue to cause environmental damage and store up long-term economic and health problems for ourselves. Our current culture and government policies are guided by concepts such as democracy, human rights and economics; but these are ultimately meaningless unless we recognise our fundamental dependence on the living part of our planet.

Even if we live in large cities, renewable resources are all around us. We import huge quantities of virgin wood pulp, yet we could meet a large proportion of our needs from the millions of tonnes of high-grade waste paper we send to landfill. London, for instance, is permeated and surrounded by woodlands that could be managed sustainably for a variety of products. Even street trees, managed as a sustainable forest, can provide a renewable energy supply. Other resources include solar energy falling on the city, sewage which is full of nutrients, and glass and plastic in the waste-stream.

Since 1994, BioRegional has been working to create sustainable products and services based on using local renewable and waste resources. Our partners include major companies like B&Q, M-real, Inveresk plc, BP and Cosmopolitan Cosmetics. We have teamed up with UK housing providers like The Peabody Trust, as well as community groups and a local prison. The products and services developed within a BioRegional framework and described in this book are working examples of thinking globally and acting locally within a market economy, and aimed at reducing our ecological footprint.

Using local renewable and waste resources wisely allows us to create communities that locally recycle many of the materials and nutrients required to support them, much like natural eco-systems. This ties in with the UK government's Proximity Principle, advocating local recycling. Local Paper for London, our closed loop office paper recycling and buy-back scheme, is one working example. Converting wastes from one activity into resources for another using the principle of 'industrial ecology' will

---

* In this Briefing we use 'BioRegional' for our branded products and services, whereas we use 'bioregional' and 'bioregional development' as generic terms.

also be an important strategy in building a sustainable future. For example we can use waste arising locally from textile production to make particle board for packing crates and furniture, as we describe in our vision for a BioRegional fibre industry.

Localisation of production and consumption is a key theme in our work. Local production cuts down on unnecessary transport, supports local economic development and re-establishes a direct link between people and their living environment. To create more locally self-sustaining communities in compact cities, we also need to reconsider the scale and organisation of our production systems. Local and regional scale (rather than global scale) technologies can form the basis of a new industrial revolution. There are many opportunities for bulky commodity products, which are expensive to transport, to be produced locally on a decentralised basis. We can set up co-ordinated networks of local producers, responsive to local needs—bioregional networks—as we have done for charcoal supply in the UK. As we move to more locally self-sustaining communities, we will see the balance of international trade moving away from low value commodities to high value added products, allowing countries to generate foreign exchange with low environmental impact as measured by our 'FEET' index.

Although we see sustainability being delivered via products and services—that is, via a sustainable market economy—we will also need strong international governance to ensure that the market operates with sustainability principles and creates a level global playing field. A market defined by sustainability criteria will automatically support more local production—where the benefits of globalisation and Comparative Advantage are balanced by those of BioRegional Advantage.

To demonstrate that it is possible to design a much more sustainable urban way of life we have developed a variety of initiatives that are described in the following chapters. One of our projects has been the design and construction of an urban eco-village in South London, Beddington Zero (fossil) Energy Development, or BedZED. Designed with architect Bill Dunster and developed in partnership with Peabody Trust, we are showing how green living is a real, attractive and affordable option for the person in the street, created by integrating energy efficiency, renewable energy and water harvesting with services like car pools and local organic food deliveries. This joined up approach to green lifestyles and the use of local sustainable building materials typifies the BioRegional approach. Bill Dunster describes bioregional development as the glue that binds BedZED together.

BedZED is the UK's largest eco-village. It is made up of 82 homes, plus office space and live-work studio apartments. The village has a mix of social housing for people on low incomes and private homes for sale at prices comparable to more conventional homes in the area. It is designed for a comfortable, yet highly resource-efficient bioregional way of life.

BedZED can trace its history back to the early 1990s, when Bill Dunster built his own prototype solar home, Hope House, in Surrey. Hope House is very well insulated, faces south and has a large conservatory on the south side to trap warmth from the sun—providing so-called passive solar heating. Bill later added a solar hot water system and photovoltaic (PV) panels. The PV panels are used to charge an electric car that provides the main form of transportation for the Dunster family and also serves as the company car for his architectural practice.

Working with environmental services engineer Chris Twinn of Arup, Bill used Hope House to create a theoretical model of a high density mixed residential and commercial development: Hope Town. With some seed funding from the World Wide Fund for Nature International, BioRegional starting working with Bill to develop this concept further and turn it into a reality. We looked at how we could use local sustainable materials, local renewable energy supply and integrate a range of services for residents to make green living an easy option. When BioRegional identified a possible site at Beddington Corner, the opportunity for a real project was born.

The site was 1.65 hectares (approximately 4 acres) of former sewage works, due to be sold by London Borough of Sutton. The land had some heavy metal contamination and so some decontamination work would be needed when developing the site. We started approaching the local community to find out their views about an eco-village. As we described the concept in more detail, the local residents' association and vicar became increasingly supportive. Bill Dunster drew up plans, incorporating ideas such as space for a nursery suggested by the local residents. The cost of the scheme was determined by quantity surveyors, Gardiner and Theobald. At this point we had a project to take to potential backers. The mix of the less conventional companies, Bill Dunster Architects and BioRegional, with well respected and established ones like Arup and G&T, proved a powerful one.

We needed a developer partner for the project. A number of house builders—big and small—were contacted and although they showed interest, felt no need to change their product from the one they were already selling to the consumer. In retrospect, we were therefore very lucky to identify the Peabody Trust as potential partners. Peabody are London's largest housing association, set up in 1870 through an endowment by American philanthropist, George Peabody. The Peabody Trust had grown to provide 19,500 homes for tenants on low incomes and developed a culture of innovation and leadership. Having invested its money wisely, the Trust had gained the financial freedom to support new ideas. Our project had now assumed the name Beddington Zero (fossil) Energy Development, to describe the aim to generate all the heat and power from renewable sources. We were invited by Peabody's development manager, Malcolm Kirk, to present the project. Peabody's development team recognised that it would take a "brave developer" to take on the scheme, but under the leadership of Dickon Robinson, Director of Development, they took on the challenge and agreed to bid for the Beddington site when it came up for sale.

Sutton Council put the site on the market in May 1998. We had a good idea of how much money conventional developers would offer for the site. We also knew Sutton Council was a leader in sustainability, being one of the first local authorities to develop a Local Agenda 21 to improve its environmental performance. Specifically, in its main planning document, the Unitary Development Plan, the council had committed itself to support a model energy-efficient housing development.

We wished to secure some financial leeway to share some of the risk we were taking in innovating, when the council would benefit in terms of meeting its own targets. We therefore pitched our offer around 10% below what we estimated would be the highest offer from a conventional developer. Although generally required to accept the highest bid in money terms, councils are free to accept a bid within 10% of the highest offer. In our bid we also placed a financial value on some the benefits such as the reduced carbon dioxide emissions and high building quality, the latter in effect being an investment in the long term future of the area. We put forward the case that if all the environmental and social factors were taken into account (i.e. internalised), our bid was effectively much higher than that which a conventional developer was offering.

Sutton Council were divided as to whether to accept the Peabody/ BioRegional offer. Sutton Council engaged economists from Aspinwall Environmental Consultants to value the environmental benefits. They placed these at around £200,000, mainly for the savings to the public purse of helping meet European Union targets for CO_2 reduction. This was in comparison to a value for the site of around £2 million. On this basis our offer was accepted, setting as far as we know a precedent for financial valuation of environmental benefits by a local authority selling land in the UK.

Sutton Council quite rightly specified environmental performance targets as part of the land sale contract and planning permission. This meant Peabody and BioRegional were bound to deliver the BedZED project in its entirety to deliver the environmental savings. There have been other cases in the UK, such as the Greenwich Millennium Village, which have failed to meet the high environmental specifications on which developers have won competitions to develop sites.

Work on site started in May 2000, and the first residents moved in to BedZED in March 2002 with full occupation achieved in July 2002. Homes range from one-bedroom flats to four-bedroom family town-houses. Both BioRegional and Bill Dunster Architects have moved offices to BedZED. Office space has been reserved by some of the residents on site. Some are being converted to live-work units suitable for home-workers including software engineers, architects and designers. There is a sports pitch and clubhouse, and a café and nursery are planned. As a mixed-use development, BedZED will never be deserted in the way that purely residential areas are during the day and commercial areas at night, lending itself to self-policing and increasing security.

The main principles which underpin BedZED are:

- reducing energy requirements to the point where renewable energy is a viable option for the entire energy supply

- designing for a lifestyle less dependent on the car, promoting car pools and electric cars

- maximising the use of local, reclaimed and recycled materials and materials with a low embodied energy (i.e. energy required in manufacture)

- reducing mains water consumption by collecting rainwater and recycling water on site

- integrating 'green lifestyle' services such as recycling and on-site composting, bulk home deliveries of groceries and local organic food deliveries to make it easy for people to chose a completely green lifestyle.

Space heating needs at BedZED are reduced by 90% compared to a typical home in the UK (e.g. one which meets the UK's 1995 Building

Regulations standards). BedZED's exceptional energy efficiency means that we haven't had to fit central heating. This is made possible by having terraces of super-insulated south-facing homes with double-glazed conservatories to provide passive solar heating. The free heat energy from the sun provides up to 30% of the heating needs of the homes.

The passive solar warmth is stored within the mass of the building (the concrete, brick and blockwork), the heavy weight construction giving the buildings a high thermal mass or thermal inertia. The buildings heat up and cool down slowly, so that heat stored from the day is released at night, and coolth from the night is carried over into the day, which avoids overheating in summer (coolth is a term meaning the opposite of warmth). The passive solar design is enhanced by super-insulating the walls with 300 mm of insulation (three times the typical amount) and also insulating underneath the foundations. The coldest days in winter are often the brightest. Therefore passive solar heating is effective in winter as well as summer, as anyone with a south-facing conservatory will have experienced. In summer the conservatory windows can be opened up to prevent over heating, in effect turning the conservatory into a balcony.

BedZED has also been constructed to be airtight to prevent heat and coolth loss, which means that if the buildings were left closed and uninhabited at the end of the summer, the interiors would take 3 months to cool down. The homes therefore carry some warmth from the summer into winter and coolth from the winter into summer. This reduction in temperature fluctuations means that it is easier to maintain a constant and comfortable internal temperature.

About 60% of the heating requirement for the homes comes from residents' body heat and waste heat from electrical appliances and cooking. The smaller homes like the one-bedroom flats will not need any extra heating at all—particularly if on an exceptionally cold night you invite a friend around. Although the homes don't have central heating, they do have back-up heating for prolonged cold spells or if elderly people or babies need a higher room temperature. Thermostatically controlled heating fins fed by the centrally placed hot water tank (heated from 'waste' heat from the combined heat and power plant on site) can trickle hot air into the home if it is needed.

Many conventional developers are reluctant to invest in higher levels of insulation because of the extra cost. But at BedZED we have increased energy efficiency to the critical point where there is no need to fit central heating. We are therefore saving around £1,500 per home, allowing us to pay for the super-insulation. This shows that it is

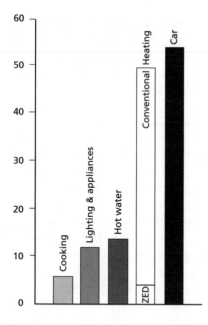

Figure 1: Energy Consumption
of a typical UK household

possible to take a conventional building budget and spend it differently to increase sustainability. As Jonathan Deans, Partner at Gardiner
and Theobald Quantity Surveyors says, "Perhaps in future, quantity
surveyors will be as much environmental cost consultants as financial
cost consultants."

However, the energy consumption of a typical UK household
involves more than just heat and power used in the home as shown in
Figure 1.

The family has an energy requirement for cooking, lighting and
appliances, hot water, heating and running the car. Although at
BedZED we have reduced the space heating requirements by 90%, we
can clearly see we cannot meet our target to reduce our ecological footprint by two-thirds unless we also tackle car use.

BedZED is therefore also designed to reduce car dependence.
Offices and workspaces offer alternatives to commuting. Our ZEDcars
car sharing pool gives residents access to a car when they need one, but
leaves them free to walk, cycle or use public transport for the majority
of their journeys. Electric car charging points powered by solar panels
offer carbon neutral motoring. Bulk home deliveries of groceries from a
local supermarket and organic food from local farms provide convenience as well as further reducing car dependence.

Ecological footprinting shows us that there is only limited value in building energy-efficient houses if we need to commute by car to and from them. We will need to think carefully and systematically about creating ways of living which design out needless commuting, transport and other forms of wasteful consumption. At the same time, the alternatives we offer must be attractive. We can sell the value of a commuting-free lifestyle as a higher quality alternative to a car dependent one; and create high quality environments for ourselves, such as we see in low car cities like Venice and Amsterdam. In Chapter 10 we further explore the green lifestyle and design features of BedZED and the imperative for sustainable cities.

BedZED, as a living and working example in the market economy, shows how a much more sustainable lifestyle is possible now—not tomorrow—in the mainstream. We feel confident that many more sustainable housing developments of this kind will spring up in the UK and elsewhere in the coming years, simply because the logic is too compelling not to be replicated.

Chapter 2

Our Planetary Means

The notion of sustainable development has been with us for about 30 years. It has variously been defined as meeting our needs without comprising the ability of future generations to meet theirs; or development that integrates economic, social and environmental needs. Efforts to define sustainability have kept a whole army of consultants busy.

However, to deal with sustainability in practical terms a range of quantitative questions need to be answered. For example: Can we grow enough trees on this planet not only to meet our increasing consumption of paper, but also provide us with wood for construction, soak up carbon dioxide and leave some wilderness areas? Can we quantify how much the planet can produce sustainably and use these figures to set targets for global consumption? Just how green do we have to be to be sustainable? If I reduce my environmental impact by buying local organic food, by how much do I also need to increase the energy efficiency of my home: by 10%, 20% or 80%? The emerging discipline of 'ecological footprinting' is helping us to find answers to these sorts of questions.

Imagine leaving planet Earth and looking back at it. We would see mountains, forests, plains, seas and lakes. We would also see clear signs of human impacts—forests cleared for agriculture, areas converted to cities. We would, like orbiting astronauts, become aware of just how small the planet is and that we are wholly dependent on it to meet our needs. This perspective of the Earth as a planet with limited space is the starting point for ecological footprinting. In August 2001 Frank Culbertson started a four-month tour of duty on the International Space Station. Upon his return he expressed his concern at the impact humanity is having on the Earth's environment. He said that he and his fellow astronauts had witnessed many signs of environmental change (see web page extract).

WWF's Living Planet Report 2000 has listed all the biologically productive land, or available biocapacity, on the planet:[1]

The total biocapacity, 12.6 billion hectares (126 million km²) accounts for 25% of the Earth's surface. The rest is desert, high mountain or deep

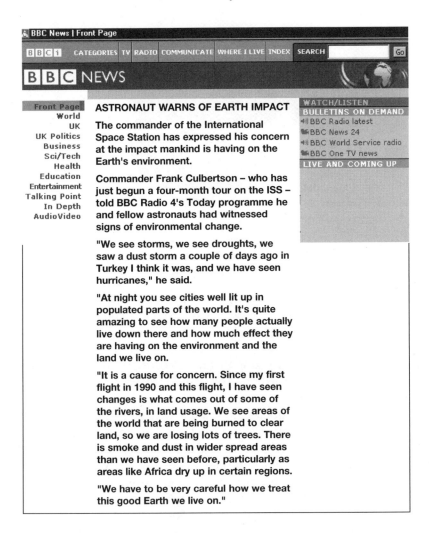

ocean which are of low biological productivity. If we put aside 10% for wildlife, we are left with 11.3 billion hectares of biologically productive land and sea from which to meet human needs. We can consider this our global budget.

The ecological footprint concept quantifies the biologically productive areas of land and sea required to meet our consumption of food, energy, materials and for absorbing our wastes. So, for example, on average, we need 25 hectares of fishing grounds for each tonne of fish we consume each year. We need 1.3 hectares of forest for each cubic

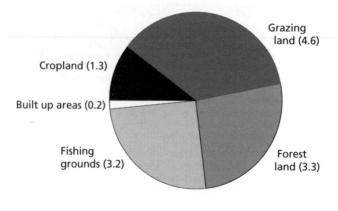

Figures in billion hectares

Figure 2: Biologically productive land

metre of timber we use annually; and 0.35 hectares of forest to absorb each tonne of carbon dioxide we release from burning fossil fuels each year. By adding up how much food, energy, building materials, paper and so on we are consuming, we can calculate how much of the earth's biocapacity is required to support an individual, a city, a country, or indeed the world's population. This last collective figure is our world ecological footprint.

Ecological footprint studies suggest that as a global society we started to exceed the biocapacity of the planet—its long-term carrying capacity—in the early 1970s (Figure 3). Currently we are consuming around 30% more than the planet can sustain. In other words, we need one and a third planet Earths to meet our current global levels of consumption. In effect we are maintaining our current lifestyles by eating into the natural reserves, or natural capital, of the planet. For example, we are doing this by losing forest areas around the world, depleting fish stocks, mining the soil of its fertility and burning fossil energy reserves to fuel our current levels of consumption.

The aggregate figure represented by the world ecological footprint hides large differences in the consumption between nations (Figure 4). Whereas the global average per capita footprint is 2.85 hectares, the smallest per capita footprints are found in Eritrea (0.35 hectares per person) and the largest in the United Arab Emirates (15.99 hectares per person).

If we divide the available biocapacity figure of 11.3 billion hectares by the 6 billion human population of the planet, we get a figure of 1.9

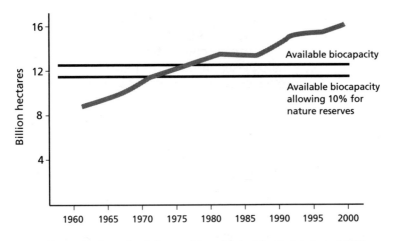

Figure 3: Growth of the world ecological footprint since 1960

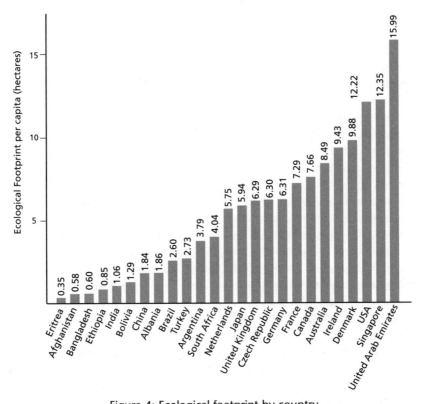

Figure 4: Ecological footprint by country

hectares (4.7 acres). This represents a per capita 'fair share' of the planet's resources. If we lived sustainably, this figure would represent, as a global average, the maximum per capita consumption.

However, the average person in the UK currently has an ecological footprint of 6.29 hectares, or three times the global per capita target. This means that if everyone on the planet consumed as much as the average UK citizen we'd need three planets to support us. If we in the UK decided to live within our fair share of the Earth's resources, we'd need to reduce our ecological footprint by two-thirds. With the average person in the USA having a footprint of 12.22 hectares, if we were all Americans, we'd need 6 planets to support us (Figure 5). Americans would, on average, need a five-sixths (85%) reduction to be sustainable.

Figure 5: If we all lived like Americans, we'd need six planets to support us.

The targets for reducing our ecological footprints may sound very challenging and politically unacceptable, especially when we are continuing to increase consumption and so are moving in the wrong direction. However, they are achievable, but significant changes are needed. We will need to move from fossil fuels to renewable energy, reduce wasteful consumption, recycle the vast majority of our waste and use virgin resources much more efficiently. We will need to reduce unnecessary transport and employ science and technology wisely. We will have to practise good housekeeping. As nations we take for granted that we should try to balance our financial budgets. In future we need to balance ecological budgets as well. One day, good housekeeping, good economics and good planet management will all be the same thing.

Our target to reduce the UK's ecological footprint by two-thirds is mirrored in a recent report by the Royal Commission on Environmental Pollution funded by the UK government.[2] The Commission investigated the need to stabilise CO_2 in the atmosphere and the possibility of reducing CO_2 levels by locking it up through planting trees. However, planting trees, even on a global scale, could only compensate for a small amount of the rising emissions. Therefore, in order to stabilise CO_2 levels, the Commission recommended that the UK should reduce fossil fuel use by 60% by 2050. This would also allow us to converge to a global per capita quota. Global quotas, based on the principle of Contraction and Convergence,[3] were seen as the only viable basis for agreeing international limits in the atmospheric CO_2.

Technological fixes alone will not allow us to reduce our ecological footprint. For instance, increases in agricultural productivity brought about by the Green Revolution have been achieved by using mineral fertilisers, themselves requiring fossil fuel inputs. As a result, modern farming has a much bigger ecological footprint per tonne of food produced than many traditional forms of agriculture and so is less efficient.[4] Another example is modern car engines, which have increased in efficiency over the years. But because there are more cars (and more big-engine cars) on our roads than ever before, any possible environmental benefit has been cancelled out. Technology will not save us from having to confront the simple issue of limits to consumption.

A small minority of academics claim that we don't face a major environmental crisis. For instance, Bjorn Lomborg[5] cites falling levels of poverty and increases in air quality in major Western cities such as London as evidence that economic growth will improve environmental conditions. However, it is apparent that we have bought many short-term gains at the cost of storing up long-term problems. We may have improved local air quality in London by installing catalytic converters in our cars, but more cars on our roads means that we are accelerating global warming. We may have mitigated a local environmental problem, but we are worsening a global one.

It is beyond reasonable doubt that global climate change is taking place. According to the Panel of over 1000 top scientists and experts on the IPCC working groups, "The rate and duration of warming of the 20th century has been much greater than in any of the previous nine centuries. . . . Emissions of greenhouse gases due to human activities continue to alter the atmosphere in ways that are expected to affect the climate. . . . Emissions of CO_2 due to fossil fuel burning are virtually certain to be the dominant influence on the trends in atmospheric CO_2 concentration during the 21st century. . . . The projected rate of warming [during the 21st century] is much larger than the observed changes during the 20th century and is very likely to be without precedent during at least the last 10,000 years."[6]

Then there are clear cases where we have put too much pressure on natural resources: for instance the collapse of cod fisheries in the North Sea. In depleting fish stocks we have reduced the natural capital of the planet and its biocapacity. So, as a global society we have fewer resources from which to live.

What has our response in the UK been, having depleted our own fish stocks? In the North Sea, quotas have been introduced to give time for stocks to start replenishing themselves. Through a World Wide Fund for Nature initiative, the Marine Stewardship Council (MSC) has been set up to certify fish as coming from well managed stocks, so that consumers can start making informed choices when buying fish. Supermarkets such as Sainsbury's have begun to stock MSC certified salmon and mackerel. But at the same time Sainsbury's has moved on to start depleting someone else's fish stocks. *The Grocer* Magazine on 17th March 2001 printed the following press release from Sainsbury's:

Cod-like fish never sold before in the UK

Sainsbury's is introducing new types of fish to the UK in a move to alleviate pressures on the Northern hemisphere's stock of cod and haddock. The supermarket is to start selling Kingklip, Panger and Kob from South Africa from mid-March. The three fish, which have never been sold in the UK before, are similar in texture and taste to cod and haddock, and will be available fresh from the fish counter. The fish are all line caught and purchased directly from day boats in Mossel Bay, South Africa. They are then tightly packed in ice and air freighted in polystyrene boxes directly to the UK, to ensure they are in the best condition when they reach the supermarket shelves. Point of sale material will suggest recipes and usages." (Sainsbury's Press Release)

The environmental impact of Sainsbury's action in this case is large and negative. If we treat Southern hemisphere fish stocks as we have done in the North, they too will eventually collapse. A local source of protein for Africans is being lost and auctioned off to a supermarket who over time will force prices down. There is also the massive increase in CO_2 from airfreighting—each tonne of freight transport by air from South Africa releases some 4.5 tonnes of CO_2 into the atmosphere. And, in this case, not only fish but also ice will be flown around the world. Thus while Sainsbury's support sustainable MSC certified fisheries, by air freighting fish as well the net effect is to make our planet even less sustainable.

Overexploiting fisheries is just one example of losing biocapacity. Globally we are losing forests and degrading valuable agricultural soils through leachates from landfill and use of pesticides and herbicides. By poisoning natural systems, we further reduce biocapacity. Yet it is clear that since we are already living above our planetary means, we simply can't afford to act in this way. We have to start doing things differently. There are alternatives: we can increase local and global biocapacity in a number of ways. We can employ farming practices that rebuild soils, or by judiciously employing techniques like combining forestry and grazing, we can get sustainable multiple yields from the same area of land, such as in permaculture systems.[7]

It is true that ecological footprinting is an emerging discipline and is not an exact science.[8] For instance, no one has yet precisely quantified the level of natural reserves we have, or how far natural processes are able to buffer global warming. However, if we wait for more precise figures it could be too late to conserve biocapacity and halt global climate change. Ecological footprinting does give us a broad indication of the challenges ahead of us. It sets the context within which we have been developing BioRegional solutions.

Chapter 3

A Sustainable Local Paper Cycle

Over the years, paper has been the focus of much attention from environmentalists. Emotive pictures of clearcut forests and fish deformed by chemicals from pulp mills have led to a high level of public awareness about the environmental impact of paper production and a drive to recycle waste paper. And yet, as Jonathon Porritt, now chairman of the UK Government's Sustainable Development Commission, commented in 1997: "If there is one industry which could be sustainable it is the paper industry, which is based on a renewable and recyclable raw material".[1] The industry has responded to criticism with some positive actions on forests and emissions, but we are consuming ever-increasing quantities of paper and are far from making our paper use sustainable.

At BioRegional we have been working on practical ways to meet our needs for paper by promoting a closed-loop approach based on locally available resources such as waste paper, wheat straw, wood and hemp. We have introduced a 'get your own back' campaign, where offices recycle their paper and buy it back, as well as pledging to reduce paper consumption. In an unusual partnership between an environmental group and the paper industry, we have developed award-winning new technology to make paper pulp on a smaller scale—the BioRegional MiniMill. Commercialising the MiniMill has led us to work in China, where the technology is most urgently needed. Because we want to ensure that these ideas and technologies are taken up by the market, and so the mainstream, we have had to continually review and develop our approach to ensure truly cost-competitive alternatives. We explore our experience in developing a sustainable paper cycle in this chapter and the next.

But if we are to propose a sustainable paper cycle, we need to address the popular myths which exist, and issues which consumers and the paper industry always raise: Isn't it better to burn paper for energy than to recycle it? For each tree cut down, two more are being planted so aren't the forests being managed sustainably? Forest cover is increasing in Europe, so aren't there enough forests to supply our paper needs?

There has been much debate about whether it is better to recycle or incinerate waste paper. In 1997 an article in *New Scientist* magazine hit the headlines: it argued that, using life cycle assessment, incineration of paper for energy was preferable to recycling.[2] This conclusion was contested by the 1998 study by Ecobilan Group, again using life cycle assessment, which suggested that recycling rather than incineration was, on balance, better.[3] Another study, sponsored by the British Newsprint Manufacturers Association, highlighted the economic benefits of recycling compared to incineration, with recycling generating ten times the income for the UK and creating three times as many jobs.[4] A life cycle assessment study produced for BioRegional by the Centre for Environmental Strategy at Surrey University[5] suggests that the best option is to recycle paper locally to reduce the environmental costs of transporting it, recycle the majority of waste paper, and incinerate only small amounts of un-recyclable paper to produce some energy for the paper mill. This scenario has the lowest environmental impact. As we shall see later, ecological footprinting, another tool to assess environmental benefit, comes down clearly in favour of recycling.

Loss of forest cover and forest quality is a second issue. Some paper companies have responded to concern about unsustainable paper production by signing up to independent environmental certification schemes such as those of the Forest Stewardship Council (FSC), but there are still plenty of horror stories. For example a report published in 2000 by the World Wide Fund for Nature (WWF) states that most of the wood consumed in Indonesia's pulp mills is still derived from clear cutting old growth Indonesian forests, with some 800,000 hectares destroyed in the last ten years for paper-making alone.[6] Friends of the Earth (FoE) found that paper produced from ancient Indonesian forests is being sold in the UK and the pulp is being purchased by seemingly responsible multi-national paper companies for use in paper mills in China.[7] Greenpeace's ten-year campaign to protect Canada's Great Bear Forest has had some success, but the last remaining temperate rainforests there are still being clearfelled to produce toilet tissue and newsprint.[8] Claims of replanting as evidence of sustainable management can be weak. Too often, diverse old growth forests are cut down and are replaced by monocultures of conifers or eucalyptus—in some cases we may not have lost forest cover, but we have lost forest quality.

Bjorn Lomborg, in his book *The Skeptical Environmentalist*, argues that environmentalists have exaggerated the scale of forest loss,[9] and maybe some have. The UN Food and Agriculture Organisation calculated a net loss of forest cover of 2.4% over the 1990s, but with many natural forests replaced by plantations.[10] The United Nations Environment Programme (UNEP) commissioned a satellite survey of world forest cover in 2001. Commenting on the findings, UNEP Executive Director Dr Klaus Toepler, stated that "short of a miraculous transformation in the attitude of people and governments, the Earth's remaining closed canopy forests and their associated biodiversity are destined to disappear in the coming decades."[11] His observation reflects Commander Cuthbertson's comments from the International Space Station noted in Chapter 2. Although we can be baffled by conflicting reports and statistics, whether from industry or NGOs, we can't ignore what we can see from satellite imaging or the eye witness accounts of astronauts.

Lomborg is mainly concerned with increased expenditure for environmental protection, such as the cost of implementing the Kyoto Treaty to reduce CO_2 emissions globally. He states quite reasonably that the costs and benefits of any investments should be compared to similar investments in all other important areas of human endeavour. Perhaps Lomborg would consider efficient and equitable use of the world's resources (such as using forests within natural limits and maintaining their biodiversity) as a good investment that would bring many benefits.

A third question is whether it is theoretically possible to meet current and predicted human demand for forest products, including paper, from sustainably managed forests; or do we simply not have space on the planet to meet our demands? Paper is of course a key product to consider, as it accounts for 42% of industrial forest use, or 20% of the world's total wood harvest,[12] the latter which includes clearing land for agriculture and firewood. This proportion is expected to grow, as world demand for paper is expanding faster than for other wood products. Paper consumption worldwide is predicted to double by 2020,[13] with new demand mainly coming from countries in Africa, Eastern Europe and Asia where per capita paper consumption is currently low.

Friends of the Earth claim that we have already reached the global limit of sustainable wood consumption. Based on these figures, FoE suggest that if we are to have equitable distribution of the global forest production, developed countries such as the UK who use more than their fair share must cut consumption of wood products by 73%.[14] This again reflects the two-thirds reduction in ecological footprint which we use as our benchmark target. However, a major study by IIED[15] concluded that there is sufficient forest to meet increased demand for paper in the future if new tree plantations are established in the tropics. But the fact is that we need our forests for more than just paper: 53% of the world wood harvest is used for fuel,[16] especially in developing countries; we need forests for timber and for soaking up carbon dioxide; we should also leave some purely for wildlife. So although we might be able to plant sufficient forests for ever increasing paper consumption, we'd have no room for anything else! Clearly we need to reduce consumption and if we take into account limited space on the planet (which is what ecological footprinting does), we need to recycle more paper as well.

It is hard, if not impossible, to come up with definitive answers to the availability of global resources. However, we are losing forests and we are increasing consumption. Sooner or later, we will run into problems. We don't necessarily need definitive answers for us to realise that it is only prudent that we consider reducing consumption and using our natural resources more wisely. In our view, developing bioregionally is simply a common sense response, allowing us safely and surely to progress towards sustainability when we don't, and never will, have all the answers.

Taking the BioRegional approach, in the early 1990s we started to investigate how we could produce more of our paper in the UK as part of a sustainable local paper cycle.[17] If production, consumption and recycling take place regionally, or at least within the UK, we will naturally become more resource-efficient, reducing our global ecological footprint. We can reap environmental benefits while supporting truly sustainable jobs and industries.

In the UK, we import 72% of the pulp and paper we consume[18] and, apart from FSC certified paper, it is still difficult to be sure that the paper we buy is produced in a sustainable way. Certainly a proportion of it, as we have seen, will be coming from completely unsustainable sources such as Indonesian old growth forests. Around 35% of paper and card is currently recycled, around half the maximum predicted wastepaper recovery rate of 72%, with most of the recycled fibre going to make newsprint and cardboard packaging.[19] With newsprint and packaging already having a large recycled content, the way of reducing our ecological footprint here is to reduce the amount we consume. For example our newspapers, especially weekend newspapers, have been getting larger, filled with junk articles; and many goods are still over-packaged.

The majority of imported pulp and paper is printing and writing paper, also known as graphics paper—which includes office paper, glossy colour magazines, catalogues and junk mail. Graphics paper is the highest quality paper and has the greatest environmental impact of

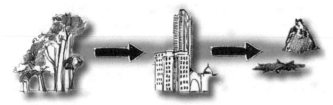

Figure 6: Forest to bin—unsustainable paper consumption in 2002

all types of paper as it requires considerable energy, water and chemicals to produce.

A staggering 97% of graphics paper and pulp used in the UK is imported, costing the UK £2.5 billion every year and requiring in the region of 1–1.5 million hectares of forest around the world to produce it. Only 5% has any recycled content at all.

We can describe our current graphics paper production and consumption pattern as one of Forest-to-Bin (Figure 6). It is made up of a long straight line stretching from forests around the globe to paper decomposing in a landfill site, or incinerated, in our communities around the UK. Although it is the highest quality paper, just 14% is recycled, mainly into toilet rolls and newspaper.

Against this background we decided to focus our efforts on graphics paper in the UK and on ways in which we could increase sustainability in its use.

We use paper very wastefully: for example sending out tonnes of junk mail (which today makes up one third of all post) and failing to photocopy double-sided. As mentioned in the case of newsprint, if we are to reduce the ecological footprint of our paper use, it will be important to reduce unnecessary and wasteful paper consumption.

We discovered that there is a great deal of fibre in the UK which we could collect for recycling or harvest from farms and local forests to make graphics paper.

The first fibre source we should consider, of course, is waste paper. In the UK in 1999, 4 million tonnes of graphics paper was landfilled or incinerated. Our research shows that there is potential to collect at least an additional 2.2 million tonnes of this for recycling in 2002,[20] in itself enough to meet a quarter of total UK demand for graphics paper. Inevitably fibres shorten and degrade in the recycling process, which means that paper can only be recycled up to five times. Therefore some virgin fibre will always be needed.

Agricultural fibres are a potential source of virgin fibre. The UK produces a surplus of 4 million tonnes of wheat straw per year. Since the ban on straw burning in the UK, the surplus straw is chopped and incorporated into the soil. In some areas this has a small positive value as a fertiliser, but in heavier soils it does not mix into the soil and forms a dense rotting mat beneath the surface—so alternative uses would be preferable. Historically, straw was a paper-making material and can be used to make a good quality graphics paper (see our description of our MiniMill technology in chapter 4).

Flax and hemp, traditional UK fibre and seed oil crops, can also be used to make the very best quality paper. The use of these fibres for paper-making becomes increasingly attractive and viable when considered in the context of a revived flax and hemp textile industry. As we explain in chapter 7, the short waste fibres from the textile industry are the perfect raw material for papermaking. Over 2,000 hectares of hemp are currently grown in the UK. It is reasonable to suggest that at least 200,000 hectares (around 4% of cropland) could be used to grow hemp and flax for textile and paper fibre, producing 2 tonnes of fibre per hectare.

In addition, wood is becoming increasingly available in the UK. As softwood plantations in Scotland, Wales and the North of England grow to maturity, an additional 4 million tonnes of wood will be available annually by 2017.[21] Although these plantations currently have a poor biodiversity value,[22] they are being improved by planting a greater

---

Potential resources available annually for graphics paper production in the UK

waste graphics paper	2.2 million tonnes
surplus wheat straw	4 million tonnes
broadleaved woodland	1 million tonnes
new forestry plantations	4 million tonnes (by 2017)
hemp & flax	200,000 tonnes
Total	11.4 million tonnes
Total paper which could be produced at 40% average yield	4.5 million tonnes

number of native species. There are also small broadleaved woodlands across the UK, which could provide around 1 million tonnes per annum.

Using the above sources of fibres we could easily produce 4-5 million tonnes of graphics paper annually, and therefore, if we chose to, be self-sufficient. Creating an inventory of locally available resources for the UK is an exercise that could be repeated in any country or region. We do not suggest that this offers the definitive solution for sustainable paper supply, but it does illustrate how even a densely populated country like the UK could be making much better use of the resources around us.

So how do we see these local fibre sources becoming incorporated into a sustainable local paper cycle for graphics paper?

Our BioRegional concept is illustrated in Figure 6, focusing on office paper use. Paper is collected for recycling from offices and taken to the local recycling mill. Here it is upgraded with 20% virgin fibre to com-

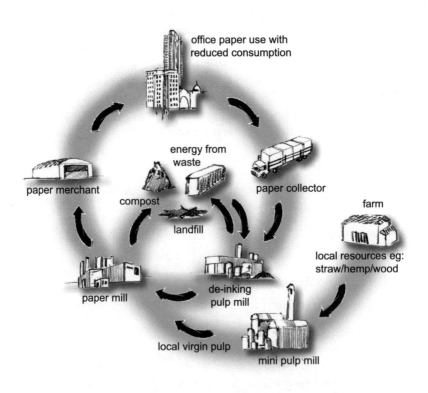

Figure 7: BioRegional sustainable local paper cycle

pensate for the fibre lost or degraded by the recycling process. The virgin fibre can be either agricultural fibre like wheat straw, hemp or flax, or wood fibre, whichever is available locally. This paper is then supplied back to the offices who use paper wisely, by photocopying double-sided for instance.

We have been working to make the sustainable local paper cycle a reality, and are testing the practical and economic feasibility of it through two initiatives: Local Paper for London and our BioRegional MiniMill technology.

In 1999, BioRegional launched Local Paper for London, a scheme for businesses in London and the South East of England. This is a closed loop scheme in which offices recycle their paper to the local mills and buy back the paper produced. We started to work with the two graphics paper recycling mills in the London area. One is a new state of the art de-inking plant in Kent, then owned by UK Paper (now by M-real), producing very high quality photocopier and general business paper under the brand name EVOLVE. The second is a smaller, long established company, Frogmore Mill, producing heavier weights and coloured recycled paper. The promotion of Local Paper for London is funded through the landfill tax credit scheme, where landfill tax is used to promote recycling and other environmental initiatives.

To promote the closed loop recycling concept and to make it easy for people to participate, we offer free advice and practical support. We contact offices, discuss their needs and any concerns they may have about quality and price. We help them set up recycling and provide them with lists of waste paper collectors and stockists of the locally recycled paper. Many companies save money on trade waste charges by segregating white office paper, which is collected free or for a nominal charge. In our scheme, offices also commit to reducing paper consumption. We provide free posters promoting tips to save paper and money ("Practise good eco-nomics"), such as using scrap for internal memos or drafts, and photocopying double-sided.

Local Paper for London has been a success and continues to grow, with over 400 organisations recycling and buying back over 2,000 tonnes of paper in mid-2002. The initiative has received the enthusiastic support of London's Mayor, Ken Livingstone. Participating organisations include large companies such as Direct Line Insurance, the House of Commons and a number of London boroughs, as well as schools and small offices. The project is changing attitudes—almost all enquirers ask us for help in dealing with their waste paper, but many are then also enlisted into reducing consumption and buying the recycled paper back. Participating in the local paper loop makes them think about the life cycle of paper, a principle which they can then apply to any product they use. As staff move to new employers they often contact us again to help set up the loop in their new office.

The waste paper collectors, such as London Recycling, have also begun to sell the local recycled paper, delivering it when they pick up the waste. This reinforces the loop concept and cuts down on one van journey in London—improving further the environmental benefit. The owners of the paper mill, M-real, are starting to consider their product as a service, contemplating providing a recycling bin labelled "Send your paper to us for cleaning". The process of moving from selling a product to selling a service, is known by some as dematerialisation. One company which has pioneered dematerialisation is the US carpet man-ufacturer Interface. As part of their business, instead of selling carpet they sell carpeting services. Interface lease the carpet, take it back when it is worn, melt it down and re-spin the carpet, ready for re-laying.[23] In this way we no longer have a waste problem.

The success of the Local Paper for London scheme has led us to start developing a sister project in Scotland, Local Paper for Scotland. However, there are a number of factors that continue to slow down the wider adoption of the concept.

There is a widely held perception that recycled paper is inferior. It is true that in the early days of recycling technology the paper produced was grey and speckled, shed dust and jammed in photocopier machines. However, the latest de-inking technology avoids these problems, and the paper produced is almost indistinguishable from virgin paper. Direct Line Insurance saved money switching to recycled paper, not only because it was cheaper than the virgin paper they were using, but also from a decrease in down time on high speed photocopiers. As Nancy Duncan, the paper buyer at Direct Line asserts, "We have had fewer jams since we changed to the recycled paper." But most printers and paper purchasers remain prejudiced. Less than 5% of graphics paper consumed in the UK has any recycled content,[24] and this is much the same around the world.

Then there is the issue of price. Recycled paper produced at the M-real mill in the UK costs the same as the non-recycled paper they make. However, the current high value of sterling means that consumers can buy imported virgin paper of a similar quality for up to 5-8% less. In addition, sometimes paper merchants increase the price of recycled paper as they know some consumers will pay more. They also offer virgin paper at below cost price in order to secure contracts knowing they can make up the difference on other stationery products. Even such a minor saving on a bill which is probably a small part of a company's total expenditure, can prevent companies from purchasing the UK produced recycled paper. The same is true for government departments and local authorities, which even if they want to, can have a difficult time convincing their bosses, the elected representatives, that they should be buying locally recycled paper. Ironically, this situation persists when government itself is spending millions each year encouraging others to recycle and buy recycled. Moreover, without a flourishing local recycling mill and a market for waste paper, all these organisations would have to spend more on waste collection.

UK industry also finds itself at a disadvantage with the introduction of the climate change levy, a tax on industrial energy consumption. The levy has increased manufacturing costs, making UK-produced paper less competitive. We then import paper made in other countries without this tax, which in some cases may be made in energy-inefficient mills, at the same time consuming more energy because we have to transport it to the UK. The net effect of the UK climate change levy on the global paper industry is then an increase in energy consumption and an increase in the ecological footprint of paper consumption in the UK. This demonstrates that such taxes are really only going to be effective if they are introduced across the world, or we have some other way of valuing the energy used in the manufacture of the products we consume. Otherwise the UK can look good in reducing energy consumption within our own borders, but in fact we are simply exporting the issue. That is not to say that we at BioRegional don't support the introduction of the climate change levy, but we might have to develop an embodied energy tax to create a level playing field between imported and domestic products in the short term. Only then can we correct this market distortion.

Another factor preventing more rapid expansion of closed loop recycling is that it is still comparatively cheap to send waste to landfill. With only a small incentive to recycle, we are not yet reaping the benefits of economies of scale in waste collection. Office waste paper collection is as yet unco-ordinated and in the hands of a multitude of small private companies. This results in a system which is not nearly as efficient as it could be, with trucks from different companies coming down the same street to pick up waste from different businesses. In our congested cities this doesn't make sense in the long term, and is expensive financially and environmentally.

Although there are barriers to wider uptake, none of them are serious or large. An increase in landfill tax from £13 to around £35/tonne will shift the finances very quickly in favour of closed loop recycling, when paper currently costs around £700/tonne. People's attitudes are also changing, and there is a greater recognition of the value of supporting recycling. Sales of recycled paper like EVOLVE are increasing year on year, and more and more companies are joining Local Paper for London. Considered in the round, closed loop recycling is a good financially viable option if savings on waste collection are used to justify buying recycled paper. In addition if staff develop a culture of reducing paper consumption, very significant cost savings can be made. Simple measures can make big savings. AT&T set the default on their office copiers and printers to double-side mode and cut paper costs by 15%.[25] Direct Line Insurance joined our Local Paper for London scheme in 1999. Finance Director Richard Houghton commented:

> Environmental performance and cost control are very important to us at Direct Line. The nice thing about this scheme is that it meets both our needs. We get savings of around 20% on our paper bill because of free pick-ups of our waste paper, and cheaper paper coming into the office as well. It also meets this closed loop concept environmentally so we believe we're saving all round.

Of course, in the short term there will always be some people who will save money by recycling to avoid waste charges, but continue to buy the cheapest virgin paper rather than recycled paper. We could foil these free loaders by directly linking low cost waste paper collection to purchase of recycled paper, providing an incentive to those who recycle in the fullest sense of the word by buying back.

In order to accurately quantify the anticipated environmental benefits, BioRegional have sponsored a doctorate student, Tony Hart, at The Centre for Environmental Strategy at Surrey University to carry out a life cycle assessment (LCA) comparing our closed local scheme to virgin imported paper.[26] The study considers the cycle as it is currently operating, but also the potential benefits which could accrue from making improvements to the scheme.

The conclusions of the LCA are that recycling and buying back graphics paper locally is definitely better for the environment than buying imported virgin paper. The whole Local Paper for London cycle requires 13% less fossil fuel energy in raw material collection, paper production and delivery than the imported virgin paper system. Recycled paper actually consumes less than half the total energy of virgin paper in manufacture, but half the energy used in virgin paper manufacture is renewable, derived from the organic material in the wood not needed for paper production. The fossil energy consumed is usually roughly equivalent. The reductions in the Local Paper for London case are due to the efficiency of the plant and to reduced transportation in the local recycle and buy-back loop. In addition, there are significant avoided burdens from diverting graphics paper from landfill. Dumping 4 million tonnes of graphics paper adds methane to the atmosphere as it decomposes in landfill sites. Methane has 21 times the global warming potential of CO_2, consequently if we include Local Paper for London's role in avoiding waste paper being sent to landfill, each tonne of recycled paper produced save the equivalent of 600kg of CO_2 when considered over a 100 year time horizon.

The M-real mill in Kent is to install a combined heat and power plant to generate energy from some of the waste fibres and sludge generated from the recycling process. This should reduce the fossil fuel energy consumption by a further 8%. The mill also plans to build a factory to use a proportion of the waste fibres and sludge to make a construction board which can be used as an alternative to plywood or plasterboard. The product has some interesting and useful acoustic and strength

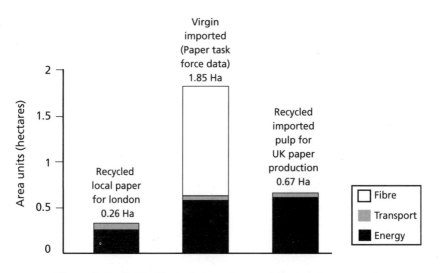

Figure 8: Ecological footprint per tonne of the local paper cycle compared to imported virgin and recycled paper

properties, which Bill Dunster Architects will be testing in a forthcoming building project.

We can envisage an ideal recycling system for a city where the paper recycling mill would be located in a combined materials and energy recycling facility. Here waste paper unsuitable for recycling through contamination, could be used to generate more of the energy for recycling, allowing the potential to reduce or eliminate the use of fossil fuels all together. This is fully explored in Hart's thesis.

The data collected for the life cycle assessment has been used to produce an ecological footprint (Figure 8) of the local paper cycle.

Considering fibre, transport and energy use, the ecological footprint of local recycled photocopier type paper is only 14% of the imported virgin equivalent.[27] If we further assume that companies joining Local Paper for London also reduce their consumption of paper by 15% by following our simple tips, the footprint totals only 12% of the virgin paper forest-to-bin alternative. With an ecological footprint saving of between 86–88%, we have a system for paper which can contribute significantly to our overall target of reducing our ecological footprint in the UK by two-thirds.

Chapter 4

BioRegional MiniMills: Scale of Technology and Local Resource Use

In the 21st century it seems that anything is possible with science and technology. We take mobile phones and the internet for granted now, but over a decade they have changed the way we live. Whilst we can't rely on technological fixes to save us from problems like global warming, the use of efficient technology to harness local renewable energy resources could make a significant impact. However, this is probably only true if we use it in the context of the bigger environmental picture. For example, photovoltaic panels might generate renewable energy from sunlight but if we eat more airfreighted, out-of-season food then the net effect on our overall energy use will be no different; indeed, our total energy consumption may increase.

E.F. Schumacher discussed the importance of scale and technology in his book *Small Is Beautiful*.[1] He developed the concept of Intermediate Technology, a people's technology that promotes decentralisation and is gentle in its use of resources. Schumacher's main focus was on labour-intensive, low cost solutions for developing countries, producing for local needs. We argue that as the world develops towards a new more equitable and sustainable state, we will need an appropriate scale of technology to allow sustainable use of local resources, and that this will be equally relevant to both developing and developed countries alike. Technologies will have to be competitive in the mainstream of our economy, not simply for craft-scale production. In this chapter we describe the development of our MiniMill paper pulping process as an example of the potential advantages of the BioRegional approach.

As we explained in the previous chapter, a supply of quality virgin pulp is a vital ingredient in a sustainable local paper cycle. However, there are no graphics pulp mills in the UK. The idea that we can use locally available and waste resources like straw, hemp or coppice wood to produce graphics quality pulp across the UK is very attractive. In

1996, as we tried to develop this into a practical proposition, we quite quickly discovered a major stumbling block. Agricultural crops and residues like straw are bulky and expensive to transport long distances. They are best pulped locally on a small scale of perhaps 10,000 to 30,000 tonnes of production per year. Yet the typical modern wood pulp mill is not considered an economic proposition below 400,000 tonnes of pulp per year, requiring an input of 1 million tonnes of raw material. This hardly allows for flexibility. The technology only works if you have a 350,000 hectare (3,500km^2) forest on your doorstep. There is an element of economy of scale, but the key factor determining size is the huge chemical and energy recovery boilers necessary to deal with the black liquor effluent produced. When pulping, cellulose fibres are extracted and the remaining plant material such as lignin and hemicellulose, together with the pulping chemicals, ends up in the effluent, which is known as black liquor. Therefore, in order to develop a BioRegional MiniMill, we needed to find and develop a way to deal with black liquor effluent on a small scale.

At a meeting in 1997 at The Royal Botanic Gardens at Kew in London, our patron, Professor Sir Ghillean Prance, then Director of Kew, addressed the paper industry. He spoke of the damage that eucalyptus pulp plantations in South America—which are promoted as sustainable—can do to the local environment. At this meeting we proposed to develop MiniMill technology. The World Wide Fund for Nature (WWF) were the first to pledge their support to our project, followed by six UK paper manufacturers and the UK Department for Trade and Industry (DTI). Environmental considerations were important, but the paper manufacturers also had sound business reasons to get involved. Despite being large companies in their own right, they are dependent on importing pulp and so are at the mercy of the multi-national pulp producers and the commodity pulp market. The price of this pulp can vary by 100% in any one year, making cash flow planning extremely difficult. The UK paper manufacturers wanted to develop MiniMill technology to be able to produce a proportion of the pulp they use at a predictable price. They also wanted to be able to make new ranges of paper products from a diversity of raw materials. For example, straw pulp gives paper a very smooth and crisp appearance, while hemp paper gives added strength and a very luxurious feel. This book is printed on paper made from 20% hemp, grown in Essex and 80% recycled paper pulp from the M-real mill in Kent.

In an unusual partnership between the paper industry and environmental organisations, we set up a company, BioRegional MiniMills (UK) Ltd, to take the project forward and develop the technology. Working with engineers AF-QPS and other experts in the field, we looked at adapting technology developed for other industries, such as in plastics, waste, textiles and food processing. Over the past 4 years, supported by the DTI, we commissioned a series of pilot-scale trials on novel small-scale black liquor treatment and more efficient pulping. The trials have proved the technical and financial feasibility of the process, demonstrating that the BioRegional MiniMill technology should use 80% less water and fewer chemicals than traditional non-wood pulp mills. It is also likely to require 50% less electrical energy than the latest conventional wood pulp mill and 90% less electrical energy than traditional non-wood pulp mills. The process will generate all its own thermal energy from the organic material in the black liquor effluent. Pulping chemicals will be recovered in a closed loop process with no damaging emissions to the environment. The paper samples produced from our pulping trials were very high quality, certainly as good as the eucalyptus pulp which is widely used in the paper industry. The capital and operating costs of the MiniMill are expected to be comparable, per unit of output, to a traditional pulp mill many times its size. So we have a strong basis on which to believe that MiniMills will be able to compete with the mega-mills. Our next step is to build a full-scale demonstration plant.

With a worsening economic climate for manufacturing in the UK, the UK paper companies were unable to commit the investment to build a demonstration plant. We had always been aware that the MiniMill technology would be of great use in China, a country that until recently used straw and other non-wood materials to make 86% of their paper.[2] Indeed paper as we know it today was invented in China by Tsai Lun, minister of agriculture in AD105, who made the first sheets from hemp fibre.

Since the mid 1990s the Chinese government has begun to close down the straw mills. They were discharging their black liquor effluent into the rivers, causing 30% of China's industrial water pollution,[3] as no small-scale technology existed to deal with it. A senior official in the Chinese government has confirmed that by the end of 2002, no mills under 30,000 tonnes of production will be allowed to operate without black liquor treatment.[4] Over seven years, 8 million tonnes of straw pulp production will have been lost in China and this has resulted in China greatly increasing imports of waste paper, wood pulp and finished paper. China has solved one environmental problem but now has two in its place. First, the straw that used to be pulped is now burned, causing air pollution and smog. Second, China is now importing wood pulp and, just like in the UK, a proportion of this pulp originates from threatened forests in Indonesia and Canada. A Friends of the Earth study reports that the biggest customer for pulp produced by Asia Pacific Resources International Holding Ltd (APRIL) who are clearcutting ancient Indonesian forests, is the multi-national UPM Kymenne who are pur-chasing the pulp for their Changsu paper mill in China.[5]

The effect of mill closures in China has even been felt in the UK. Due to China's increased demand for waste paper for recycling, the price of waste paper to UK recycling mills has increased for those that do not have a firm contract with waste collectors, leading to the temporary clo-sure of one recycling mill in Scotland.

China, like the UK, does not have a sufficient forest resource to meet its needs for paper. Indeed, following catastrophic floods in the Yangtze river basin in 1998 which were blamed on deforestation, the govern-ment banned logging in the area. Although tree plantations are being planned which will produce enough wood to make four million tonnes of pulp for use in paper production, they will take years to mature.

During 2000 we decided to explore possible interest in the MiniMill in China. A Chinese trade advisor, David Wei, was engaged to carry out a market study and he identified three potential partners. With his help we found a site for the first mill in Jiangsu province near Shanghai. The mill had been producing 15,000 tonnes per year of straw paper, but the pulping was stopped in 1999 as it was the largest source of pollution in the city. Most of the workers have been laid off and the mill now only makes small quantities of paper from imported wood pulp, which costs twice as much as the straw used to. With our main engineering part-ners, AF-QPS, we have produced a feasibility study for a BioRegional MiniMill for the site, gained the approval of the local environment agency, and met with several potential investors. The necessary invest-ment to build the first mill is now being put in place and we hope that the first mill will be built by 2004. Government officials have pledged their support to the first mill and will use their networks to inform paper companies across China when the new technology is proven and gen-erally available.

Although we didn't plan to launch the MiniMill in China, we are pleased to have started working in the country. It is clear to anyone vis-iting China, especially Shanghai, that this is where that the greatest eco-nomic growth in the world is taking place. China's paper consumption in 2000 was 36 million tonnes[6] or 11% of global paper consumption, but is predicted to double by 2010. Put another way, if China's paper consumption increased from their current level of 30 kg per person per year to the UK level of 200 kg per person, China would need 264 million tonnes of paper annually or 80% of current global paper production. What hope would there be for the world's forests then?

Appropriate technology in the form of MiniMills will allow China to make use of its straw and other agricultural residues once again. It will also reduce transport and pollution. The bulky raw material can all be sourced from within a 50km radius to make pulp and paper primarily for local use. The MiniMill is adaptable, and can be used to pulp wood if that became increasingly available. If the use of virgin pulp is complemented by local paper recycling, then we can establish a complete sustainable local paper cycle as illustrated in Figure 7 in the previous chapter. This could be replicated in any developed or developing country, using whatever sustainable fibre resource is available locally. Indeed we have already received enquiries from all over the world.

In line with our original aims, we hope to also see MiniMills established in the UK. In addition to the benefit of reducing pressure on the world's forests, if we used just 1 million tonnes of the UK's waste straw to make paper we could generate £250 million worth of paper product and create over a thousand jobs in rural areas and in manufacturing.

We hope that BioRegional MiniMills Ltd will become a successful company. The BioRegional group strategy is that as projects spin off successful companies, these companies covenant back a proportion of their profits to the central charity to develop further projects. The BioRegional charity is a major shareholder in the BioRegional MiniMill company so some of the profits from the sale of MiniMills will be used to fund the development of more BioRegional solutions in China and the UK.

Appropriate technology like the MiniMill can only play one part in the creation of a sustainable local production of paper. The right attitude is also needed. MiniMills might be able to produce paper in the UK and China from an otherwise waste material, but if paper consumption in the two countries grows unchecked and paper is wasted instead of being recycled, we can never be sustainable. We will all have to be wiser than we have been in the past if we are going to develop sustainably and live off the resources of one planet.

There are other examples where BioRegional projects are applying appropriate scale technology to harness local sustainable resources, developing market based solutions to reduce our ecological footprint.

These include the small-scale combined heat and power plant at BedZED, powered by tree surgery waste, and the photovoltaic panels incorporated into the conservatories of the homes, which convert solar energy into power for up to 40 electric cars. BioRegional are introducing regional-scale charcoal production technology, cleaner and more efficient than current kilns, to produce barbecue charcoal locally as an alternative to imports. Our hemp for textiles project has linked to Australian Fibrenova technology, which aims to produce hemp textiles at a price competitive to cotton.

In order to make use of local resources and allow a bioregional form of development to become part of the mainstream, we will need many more efficient, small-scale technologies. It may take money, time and effort to achieve, but, as we say throughout this book, it's not rocket science.

Chapter 5

Charcoal, Butterflies
and International Trade

With the growth of the global market, businesses operating in forestry, agriculture and manufacturing have become ever larger and more centralised in order to benefit from economies of scale. Likewise, the retail sector is increasingly dominated by a handful of national chains. Up against this gigantic machine it may seem hard to make alternative, local supply structures work. In this chapter we describe how we have created a BioRegional network of producers supplying local barbecue charcoal to outlets of national retailer B&Q across the UK. Our network combines the benefits of local supply with national co-ordination and marketing. We are convinced that this form of industrial organisation—network production—could be applied to a range of industries, making local production part of the mainstream. We recognise that our local charcoal scheme is substituting for imports from developing countries and we explore the implications in terms of international development. Our suggestion is that international trade should be geared primarily towards high value products—allowing countries to generate foreign exchange whilst reducing any contribution to global warming from long-distance transport.

Although the UK has a very good tree-growing climate, we import the bulk of our wood products. Neglected woodlands in the UK should be a local source of sustainable wood. In the past, many of the UK's woodlands were managed for wood using techniques such as coppicing[1]—an ancient management technique in which trees are cut back to a stump and allowed to send up a number of new shoots. Cutting back these trees every 7 to 30 years provides a sustainable harvest of wood, creating what are known as semi-natural woodlands. The harvested wood can be used for a wide range of products including furniture, fencing, firewood and charcoal.

Coppicing creates woodlands with areas of sun and shade, offering a diversity of habitats for wildlife. In the first few years after cutting, woodland flowers such as violets and vetches flourish, stimulated by the light and warmth. These flowers in turn support a variety of insects, notably woodland butterflies. As coppices grow into thickets they become home to a second generation of species such as nightingales, turtle doves and dormice. Mature coppices become suitable for a third generation of species, such as flycatchers. Areas within a coppice which are left untouched, such as on steep banks, become overgrown and dark, forming habitats for species which thrive in deep shade including ferns, mosses and beetles. So, although it may seem odd and counter-intuitive at first, harvesting wood, if done in the right way, can be positively beneficial for wildlife and increase biodiversity. Some species in the UK are now dependent on coppicing to maintain viable populations. The beautiful pearl-bordered fritillary, the UK's fastest declining butterfly species, is one example of a species that has become locally extinct as coppice woodlands have fallen into neglect.

Why is coppicing so good for wildlife? In effect coppicing mimics the natural woodland process of 'gap formation', where old trees die and fall naturally or are blown over by gales, leaving sunny areas where young trees can take hold. Coppicing has other benefits. Growth in woodland clearings is very vigorous and so harvesting wood can increase not only its biodiversity but also its biological productivity. Coppicing therefore is one form of productive land management that is sustainable in the widest definition of the term.

Over the past 20 years, conservation groups such as Butterfly Conservation and The Wildlife Trusts have been promoting coppicing in nature reserves purely for its biodiversity benefits. However, being dependent on volunteers or grant funding to pay for coppicing, conservation bodies can only manage small areas. If we could make coppicing financially viable, we would encourage a much wider uptake of coppicing, creating local rural employment as well as helping threatened species.

In the mid 1990s, supported by WWF and the UK Government's Countryside Commission, we started looking at expanding markets for coppiced wood. One potential market we identified was barbecue charcoal. Most of the 60,000 tonnes we consume in the UK is imported, mainly from tropical forests. Two-thirds of this charcoal is used for barbecue and has a wholesale value of around £20 million per annum. At first we were most concerned with charcoal from clearance of mangrove swamps. These are important habitats and globally around half of them have been eradicated over the past 20 years. Yet they are the important nursery grounds for many marine fish. Mangroves also prevent silt flowing from estuaries out into sea, and in some cases mangrove clearance has suffocated coral reefs. We are now increasingly concerned about cheap charcoal being imported from the clearance of forests in West Africa.

The UK has a very long history of charcoal-burning, dating back to the Bronze Age,[2] but it had almost completely died out by the 1980s. However, following the Great Storm in 1987, a small revival of the industry got under way, as foresters and tree surgeons experimented with products they could make from the huge surplus of wind-blown wood. Using simple steel ring kilns, it was not difficult to get started in production again and a small but ready market was found selling barbecue charcoal to local grocery stores and garden centres.

In the UK, national retailer chains control at least 70% of the charcoal market. Individually, the small-scale UK charcoal burners were making only a few tonnes of charcoal per year and could not access the national market which required hundreds of tonnes of charcoal. The big retailers are naturally reluctant to take on large numbers of small producers because it increases their administration costs. If anything, retailers are under pressure to reduce the number of their suppliers. Furthermore, small producers are often unfamiliar with the demanding conditions of supply required by national retailers. They require their suppliers to receive orders electronically, such as through the TradaNet system. The cost of installing and maintaining the electronic ordering software is around £1,500 per year, which is not viable for small producers. Then there are the costs of barcoding the product and providing the marketing support which big retailers demand. So at first sight, local producers and national retailers don't fit comfortably with each other.

BioRegional started working with the British Charcoal Group—an informal group bringing together the UK's charcoal burners—to implement a solution. We proposed supplying the big retailers via a national network of local producers. We set up a company called BioRegional Charcoal Company Ltd to coordinate the network and to supply barcoded bags to burners, who would deliver charcoal directly to the local outlet of the national retailer. BioRegional Charcoal Ltd set quality and environmental standards to ensure that the network operated effectively. BioRegional Charcoal also subscribed to TradaNet and could therefore download orders from retail stores anywhere in the country, passing them on to the local charcoal burner when deliveries were required. Being small independent producers, the possibility of gaps in supply existed, for example if a charcoal burner's vehicle broke down. However, the network could cope with this by BioRegional simply ringing up the next nearest charcoal burner to back up the delivery.

The concept is that the network behaves as a single supplier. Retailers negotiate prices and quality standards with one person—a BioRegional representative. The product, though locally produced, is quality controlled and carries a common barcode, allowing the big retailer and small local producer to work together.

Our network production model has worked effectively for 6 years, showing local production for local needs in action in the mainstream market. Our main customer is B&Q, the market leader in the DIY (home improvements) sector, a company which has pioneered the introduction of sustainability criteria into its product range. Although more expensive than imported charcoal, BioRegional charcoal sells well, being very high quality and gaining loyalty from some customers because of the story behind the product. The network has grown from 1995, when 30 B&Q stores were supplied from 15 charcoal producers, to supplying all 300 B&Q stores with 40 suppliers from Scotland to Cornwall (Figure 9). In 1997, the scheme was expanded to include local firewood and kindling.

With a turnover of around £350,000 per year and each £25,000 of turnover supporting the equivalent of one full-time person, the company is currently supporting 14 rural jobs. We appreciate that this is not large in the scheme of things, but it is a working example in an otherwise declin-

Figure 9: Map of the
BioRegional Charcoal Network
of local producers

ing sector. The success of the BioRegional network production model has wider economic significance in the context of falling incomes in rural areas of the UK. Much of the income generated by the charcoal burners is, in turn, spent in the local economy, such as rural shops, helping to support rural communities and thus multiplying the benefits. We will return to the importance of this local 'multiplier' in chapter 9.

Each tonne of local charcoal and each 6 tonnes of firewood sold supports 1 hectare of coppice woodland in long-term management. So BioRegional Charcoal Company is supporting around 300 hectares of woodland. Working with Butterfly Conservation in the south of England, we have also targeted some woodlands supporting the last remaining populations of pearl bordered fritillary in the area, helping them maintain a foothold for the future.

A research project by a University of London student, Antony Hart, quantified the transport energy required to import charcoal, say from South Africa, compared to delivering locally.[3] We can use Hart's figures to calculate the quantity of CO_2 released per 3 kg bag of charcoal for each stage of its journey to the retailer's shelf:

S. African plantation		
↓	400 km	0.23 kg CO_2
Port in South Africa		
↓	9,540 km	0.77 kg CO_2
Port in UK		
↓	320 km	0.07 kg CO_2
Importer's Warehouse		
↓	240 km	0.07 kg CO_2
Retailer's Warehouse		
↓	410 km	0.18 kg CO_2
Retail outlet		
Total	10,910 km	1.32 kg CO_2

Therefore 1.32 kg of CO_2 is emitted for each 3 kg bag of charcoal we import. It is worth noting that the long sea journey which accounts for

almost all of the total distance actually accounts for only about half of the CO_2 released. Half of the CO_2 released arises from van and lorry journeys to and from the ports.

We have compared the transport CO_2 emissions of imported charcoal to our own local charcoal. In the case of local charcoal there is only one short journey from the production site to the retail outlet, an average trip in a small van or Land Rover of 48 km. Although these vehicles are a lot less energy-efficient at deliveries than bulk haulage lorries, the transport distances are very much shorter.

UK woodland production site

↓   48 km   0.13 kg CO_2

Retail outlet

The CO_2 released per 3 kg bag of BioRegional local charcoal is 0.13 kg. Therefore through our local supply network we have reduced the CO_2 footprint of transporting charcoal by 90%. This is more than our rule-of-thumb two-thirds reduction target we set ourselves in Chapter 2 to achieve sustainability in the UK.

However we can reduce the ecological footprint even further. The very small-scale ring kilns currently employed by our charcoal burners are not very efficient. Working with some of our current suppliers and bringing in engineering company partners, we are planning to expand with a UK network of regional-scale charcoal plants, each producing 2,000 tonnes of charcoal per year, each supporting 15 jobs and supplying around 5% of the UK barbecue market. Regional-scale production will allow us to decrease production costs so that we can compete on price with imported charcoal. Increasing sales volumes will still allow delivery distances to be kept within a 50 km radius, but using lorries rather than vans will allow us to reduce the CO_2 footprint to under 0.05 kg per 3 kg bag of charcoal—twenty times better than imported charcoal. Using this scale of technology also enables us to recover waste heat to pre-dry wood, so that we will only require four rather than six tonnes of wood to make each tonne of charcoal. We can therefore reduce the woodland component of our charcoal's ecological footprint by a third. We anticipate that even after pre-drying wood we will have surplus heat which could be used for growing food in polytunnels, for other industrial processes or for hot water to homes.

In theory, the UK could become self-sufficient in charcoal. The government agency responsible for biodiversity, English Nature, would like to return 70,000 hectares of England's 200,000 hectares of woodland into coppice management, but this can only happen if this land can be brought into economically viable use. Using regional-scale charcoal plants, we would need around 35,000 hectares (50% of the English Nature target) for the whole of the UK to become self-sufficient in charcoal. Put the other way around, we could meet 50% of English Nature's biodiversity targets by producing all our own barbecue charcoal. If English Nature had to meet this target by grant funding coppicing, it would cost at least £2.3 million per year. We would also create around 300 rural jobs, when the UK rural economy has suffered terribly in the past few years.

Using the BioRegional Charcoal Network model, we can envisage an international network of regional charcoal plants that supply locally produced charcoal worldwide. B&Q, for example, now have successful stores in Shanghai and we could set up production of charcoal in China to supply these stores.

We are often asked if promoting greater UK charcoal production is putting producers in developing countries out of business. Aren't we removing one of the few sources of cash income they have? How will people be able to raise their standard of living to help them to protect their environment if they don't have an opportunity to generate foreign exchange?

But there is much evidence that trading in commodities rarely brings real benefits to poor countries. A United Nations report has shown that the 48 poorest countries have got poorer with trade liberalisation.[4] Of the £2.80 retail price for imported charcoal, less than £0.10 goes to the charcoal burner. Indeed in some countries, such as Brazil, there are areas where charcoal burners are indentured to the charcoal merchants and child labour is commonplace. When we import products without the comfort of Fair Trade certificates, we just cannot be sure we are doing the best for these countries.

We must also not forget that we simply have no option but to address sustainability. It is certainly not sustainable to transport charcoal from the southern hemisphere to the UK. It releases 10 times more CO_2 than local charcoal, and therefore has a 10 times greater contribution to global warming. As the Red Cross, for example, rightly points out, the poorest countries in the world are the ones that will be least able to deal with the consequences of environmental damage and effects of climate change. The UN has predicted that crop yields in the tropics could tumble by as much as 10% for every one degree centigrade rise in global tempera-ture.[5] Environmental sustainability should be non-negotiable.

Although we advocate more locally sustainable and self-sufficient form of development (that is, bioregional development) in both indus-trialised and developing worlds, developing countries will need access to foreign exchange to some extent. We therefore need to foster inter-national trade which has minimal impact on the environment. In order to gain an insight into this issue it is useful to develop sustainable inter-national trade indices. Here we propose one such index: Foreign Exchange Earnings per Transport tonne of CO_2, or the FEET index. As its name suggests, we simply divide the amount of foreign exchange earned by the CO_2 released by transporting the product to the country of sale.

The different modes of transport release different quantities of CO_2. This is normally quoted as grammes of CO_2 required to transport one tonne of freight one kilometre—or grammes CO_2/ tonne-kilometre. Table 1 below is reported in Sustain's 'Eating Oil' report, with figures collected from various sources including the UK government.[6]

We can use these CO_2 emissions figures to calculate FEET indices to guide us in deciding how we might support sustainable development in South Africa (Table 2).

The 2002 export price for South African charcoal was £319 per tonne, or 5,740 Rand. Transporting each tonne of charcoal from South Africa by road and sea to the UK port releases 0.33 tonnes of CO_2. The FEET score for charcoal is therefore 17,390 Rand/tonne CO_2. Charcoal

Mode	Description	Emissions (grammes CO_2/ tonne-kilometre
AIR	Short-haul Long-haul	1580 570
ROAD	Van Medium Truck Large Truck	97 85 63
SHIP	Roll-on/Roll-off Bulk	40 10

Table 1: CO_2 emissions per tonne-kilometre travelled
for different modes of transport

Product	Export price	Rand/ tonne product	Mode of Transport	Transport CO_2 per tonne of product	FEET INDEX (Rand per tonne CO_2)	FEET rating
Charcoal	£319/tonne	5,740	Road-Sea	0.33	17,390	Low
Apples	£554/tonne	9,970	Road-Sea	0.30	33,233	Medium
Grapes	£2.80/kg	50,400	Road-Air	6.00	8,400	Very low
Good quality wine	£3/bottle	45,000	Road-Sea	0.30	150,000	High
Baygen Radios	360 rand/radio	163,800	Road-Sea	0.30	546,000	Very high
Tourism	see text		Return Air	see text	6,720	Very low
Software	see text		Electronic	see text	extremely high	Extremely high

Table 2: FEET Indices—Foreign Exchange Earnings per Transport tonne of CO2

is a low value, bulky product that is expensive to transport, and so is a low FEET product.

We have estimated the FEET score for a number of products—from air-freighted grapes to the famous wind-up Baygen (Free Play) radios, which until recently were manufactured in South Africa. We have also estimated the FEET index for tourism. In this case the CO2 is released by the tourist flying return from London to Johannesburg, some 1.25 tonnes of CO2 for the return journey. Staying for 14 days and spending 600 Rand per day,[7] the tourist will generate 6,720 Rand/tonne CO2. The excellent wines grown in South Africa are a high FEET product, but are lower than they could be as the wine is transported in bottles. If we were to transport wine in bulk containers and bottle it in the UK, the FEET index could be almost doubled.

South Africa has a small software industry. Software is written in South Africa and sent electronically to wherever in the world it is required. Countries like India are developing their software industries very rapidly (indeed far more effectively than when India's major foreign exchange earners were commodities like cotton and tea). The foreign exchange earnings from software are very high and the environmental cost of transporting the product is negligible. The FEET score will be very high indeed.

The range of income per unit of CO2 ranges from 6,720 Rand for tourism up to 150,000 for wine, 546,000 Rand for radios, and even more for software. In terms of sustainable international development we should focus on international trade in medium and high FEET products. Development issues are complex. The wind-up radios are now made in China because of low productivity in South Africa, emphasising the societal issues which exist and cannot be changed overnight. On the other hand, the South African wine industry is flourishing. What we do need to do is to work together so that countries like South Africa can gear exports to products with high foreign exchange earnings and low environmental impact. That doesn't mean that South Africa shouldn't have a charcoal industry. Far from it. We would like to set up a BioRegional charcoal network in South Africa to supply charcoal to South Africans on a local basis. We would like to see South Africans, like us in the UK, have the opportunity for a diversity of employment from rural work to manufacturing, with bulky, transport intensive products being produced and supplied locally wherever possible.

South Africa is well known here in the UK for its exports of fresh fruit and vegetables. However, many black South Africans are not even aware that the country exports so much food. A community leader from Soweto, Mandla Mentoor, visited our BedZED eco-village in south London in March 2002. We took him to see a local supermarket, displaying produce from all over the world including South Africa. He had this to say:

> I am very shocked to see so much South African fruit and vegetables on British shelves when we are reeling with hunger back home and being forced to eat genetically modified foodstuffs. If there is so much money coming from exports, where is it? We don't see the money or the vegetables.

It is clear from Mandla's comment that there isn't widespread local support for exporting basic commodities from developing countries. It often does not earn enough money to compensate for loss of food growing potential, particularly when countries like South Africa have to compete against artificially low prices caused by European Union subsidies for its own farmers. It is in effect stealing from the poor and long term, countries like South Africa will be richer for not following a development route that makes them heavily dependent on commodity markets.

Global markets and subsidy structures have created a very perverse situation. Exporting apples from South Africa removes a source of food from Africa at the same time as creating problems for farmers in the UK who have been forced to grub up their orchards. On top of this, we increase CO_2 in the atmosphere. Similarly, South African charcoal undermines our local coppice woodland industries. In becoming more regionally self-sufficient we can promote a diverse regional economy in countries like South Africa and give our coppice woodland industry a chance to flourish. In the UK we could support jobs in our countryside once again and see butterflies return to our woodlands. It is a mutually respectful form of development: each country as far as possible meeting its own everyday needs from local resources and only trading high value products internationally, resulting in high earnings and low environmental impact.

Chapter 6

Local Food, Lower Risk

In the previous chapter we touched on the impacts of international trade in food. In the UK, one third of our individual ecological footprint arises from providing us with food and transporting it to us.[1] On top of this are the additional environmental costs of packaging. 'Food miles', the distance food is transported, particularly the huge increase in air-freight, has made our food supply system very inefficient environmentally. We can greatly reduce our ecological footprint by promoting local food networks and by returning fertility to local farms by recycling compost and sewage nutrients, creating a closed loop cycle. Our research has shown that the network production system we have put in place for the local charcoal supply could also work very effectively for supplying a range of local and regional food to supermarkets.

The environmental and social costs of our current food supply industry are well documented, for example in reports such as 'Eating Oil'[2] and 'The Living Land'.[3] The distances food travels before reaching our plates has increased dramatically. Sea freight doubled between 1965 and 1999. Imports and exports of agricultural and food products by road increased by 90% between 1989 and 1999, and the distance that these products are moved increased by 51%. UK government figures show that road haulage as a whole accounts for 7% of the UK's CO_2 emissions and is the fastest growing source of greenhouse gases. Agricultural and food freight accounted for 30% of all road freight in 1999: a total of 47 billion tonne-kilometres.

The growth in air freight of food is most worrying (Figure 10). Between 1980 and 1998 the air freighting of fruit and vegetables more than trebled, and more recently air freighting of fresh fish has become increasingly common.

In a project supported by The Countryside Agency and the World Wide Fund for Nature, we investigated the scope for BioRegional food production networks.[4] The research was carried out in partnership with David Hughes, Sainsbury Professor of Food Marketing at Imperial

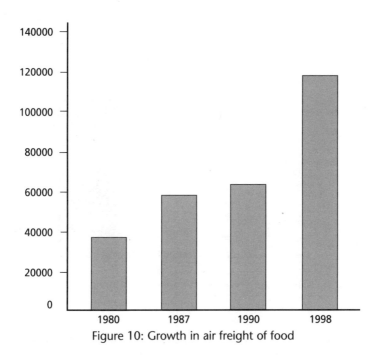

Figure 10: Growth in air freight of food

College, London. In assessing the possible benefits we looked at the case of a highly successful strawberry co-operative called Kentish Gardens (KG). Although originating in Kent, KG had grown to over 50 producers across the UK supplying 40% of the UK strawberry market. We traced KG strawberries from farm to supermarket outlets, via the supermarket retail distribution centres, and estimated the CO_2 released. When KG deliver strawberries from a farm in Kent (close to London) to a Sainsbury's supermarket in London, each tonne of strawberries releases 17kg of CO_2. When strawberries are delivered from Kent to Scotland, 145 kg of CO_2 is released. Outside the UK strawberry season, under pressure from the supermarkets to supply strawberries all the year round, KG also airfreight in strawberries from Israel. In this case each tonne of strawberries releases 4.6 tonnes of CO_2 (Figure 11).

We can relate the effect of air freighting strawberries to an individual's ecological footprint. Eating 1 kg of air freighted Israeli strawberries per year—one large punnet—has a transport footprint of 0.0016 hectares and therefore will use up almost 0.1% (one thousandth) of our target 1.9 ha ecological footprint. In a sustainable world our footprint budget will equal our financial budget. Therefore with the average UK

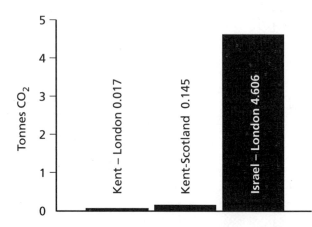

Figure 11: CO₂ emissions from the transport of 1 tonne of strawberries

salary of around £25,000 per year, in a sustainable world we would be paying one thousandth of this, or around £25, for 1 kg of air-freighted strawberries.

Reducing food miles can have a significant effect on reducing our contribution to global warming. Farmers' markets, where local farmers sell direct to the public, have proved very popular. The first one opened in Bath in October 1997, and by the end of 2000 there were over 300 farmers' markets nationwide. There are also an increasing number of so-called box schemes, where locally grown fresh produce is delivered by farmers or distributors direct to the homes of customers. Home deliveries, whether via box schemes (or from supermarkets), also offer the chance to reduce car dependence as part of a coherent strategy, such as we describe for BedZED in chapter 10.

However, the national supermarket chains dominate the food retail sector, selling over 80% of all food consumed in the UK. Supermarkets have become so enormously successful because they offer convenience and a wide range of products. Although the trend has been for supermarkets to increase food miles, there are ways in which they can support a bioregional food supply structure and increase sustainability.

The current supermarket food distribution model is based on retail distribution centres (RDCs), regionally based depots that deliver to

BedZED, the UK's largest eco-village, with 82 homes and office space for 200.
photo: Raf Makda

Full-length south-facing conservatories, combined with super-insulation and thermally massive construction, reduce the heating requirement of BedZED homes to 10% of a typical UK home.
photo: Raf Makda

BedZED, winner of the London Evening Standard Lifestyle Award 2002
—affordable and attractive green living.

The car-sharing club is an integral part of green living at BedZED.

Local Paper for London—BioRegional's recycle and buy-back office paper loop. Holding the 'get your own back' boomerang at the launch event are, from left, Jean-Paul Jeanrenaud, WWF; Alex Cordiner, Director, Shanks Waste; Sue Riddlestone, Bioregional; and Mayor Ken Livingstone.

The lavender harvester built from reclaimed agricultural machinery by a team of engineers from Cranfield University led by Dr James Brighton (left), an advisor to TV's Scrapheap Challenge. Also pictured centre and right, Paul Westrupp of HM Prison Downview and Pooran Desai of BioRegional.

BioRegional MiniMill—computer-generated view of the pulping section. *photo: AF-QPS*

Since the recent enclosure of old, polluting straw mills in China, straw is burned, causing smog.

The Bapabiozo brothers from Benin, sold to work as slaves in the cotton plantations of Ghana for £10 each.

Volunteers resting during BioRegional's hemp for textiles harvest in Kent, August 1994.

Hemp jacket by Katharine Hamnett, made using fabric produced in the BioRegional trial, the first UK-grown pure hemp fabric for generations.

Coppicing opens up the woodland floor, allowing flora and fauna such as bluebells and butterflies to flourish, adding amenity value to our woodlands. There is more than enough coppice woodland in the UK to meet all our needs for barbecue charcoal as well as a range of other products.

Making barbecue charcoal from local coppice wood at Croydon. From left: Adrian Eaton and Tony Button of BioRegional.

Alan Stewart, Garden Centre Manager of B&Q Sutton. B&Q stock locally produced charcoal supplied through the BioRegional Network.

Urban Forestry, managing street trees and local woodlands for useful wood products. Using the mobile mini sawmill, pictured from left: Tony Staples, Transco; Simon Levy, Bioregional.

Ecological footprints for UK lifestyle in hectares per person

Based on a 4-person household

		Car mileage	Car ownership (manufacture, maintenance, road infrastructure)	Public Transport	Air travel	Electricity & gas	Water	Domestic Waste	Office Footprint (energy & paper)	Food (including tranport but not packaging)	Overall eco footprint
Typical UK lifestyle	Owns car, Holidays by plane every year, Recycles 11%, Eats out-of-season highly packaged, imported food	0.90	0.41	0.00	0.30	0.45	0.002	1.70	0.80	1.63	6.19
		10,000 km/yr				22,500 kWh, electric & gas	140 litres/day		non renewable energy and virgin paper		
BedZED with conventional lifestyle	Owns car and commutes to work by public transport, Holidays by plane every year, Recycles 60%, Moderate meat eater & some imported food	0.45	0.32	0.30	0.30	0.10	0.001	1.02	0.80	1.06	4.36
		5,000 km/yr		4,000 km/yr		waste wood CHP including credit for landfill diversion	91 litres/day		non renewable energy and virgin paper		
BedZED ideal	Lives and works at BedZED, Recycles office paper, No car (member of ZEDcars club), Holiday by plane every 2 years, Recycles 80% at home, Low meat diet with local fresh food	0.09	0.04	0.30	0.15	0.10	0.001	0.34	0.16	0.72	1.90
		1,000 km/yr	20 people per club car	4000 km/yr		waste wood CHP including credit for landfill diversion wood CHP	91 litres/day		joins closed loop office paper scheme		
Global average											2.40
Global available	Leaving 10% of bioproductive land for wildlife										1.90

stores in their region on a daily basis. Farmers generally deliver fresh produce to processors or packhouses, from where the produce is delivered to RDCs nationwide. Rationalisation of food distribution has been driven by supermarket demand to reduce administrative costs, move to fewer numbers of larger producers, and exploit economies of scale. This has tended to force farmers to concentrate production for one or two supermarket chains and distribute to RDCs around the country.

However, the strawberry cooperative KG is a very good example of a network of producers delivering locally via the supermarket regional distribution system. Freshness is vital for strawberries. and so the network system has been developed primarily to reduce time between the farm and the retail shelf, rather than to reduce CO_2 emissions. The strawberry network keeps transport distances to a minimum, saving haulage costs, and KG only deliver between regions to back up supply when short in one area. The system is very similar to BioRegional's Local Charcoal network.

In our research contacting farmers and packhouse operators, they agreed that if the supermarkets asked for local and regional distribution, they would be able to form networks to deliver on this basis. Although farming has become increasingly specialised into regions, there is regional spread in growing of a variety of fruit and vegetables including carrots, soft fruit, top fruit (like apples), organic onions, potatoes, tomatoes and lettuce. The growing of brassicas, legumes and conventionally grown onions is concentrated heavily in one region; in part this is due to historical reasons rather than farming ones. If we want more local food, farmers can return to more mixed farming practices.

We can therefore envisage a more sustainable food distribution system (Figure 12) to supermarkets based on national networks of local and regional farms. The farmers would deliver to regional packhouses where the produce can be cleaned, graded and packed. From the regional packhouses, produce could then be delivered to the regional RDCs delivering to the supermarkets in the region. The farmers we spoke to were already suppliers to supermarkets, so this form of supply to supermarkets, though different, need not be dependent on smaller farms or involve less efficient ones.

The scope for promoting local food is vast, and the 2002 UK Policy Commission on the Future of Farming and Food sees local food playing a much more important role in the future.[5] To quote from the Commission's report: "We expect local food will enter the mainstream in the next few years. . . . We have heard from several supermarkets that they see local food as the next major development in food retailing."

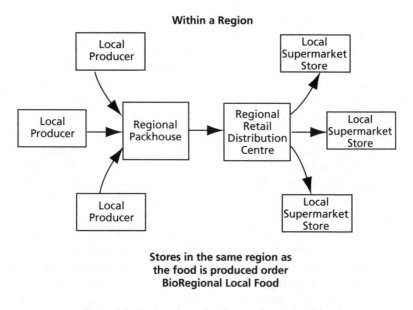

Within a Region

Local Producer

Local Supermarket Store

Local Producer → Regional Packhouse → Regional Retail Distribution Centre → Local Supermarket Store

Local Producer

Local Supermarket Store

Stores in the same region as the food is produced order BioRegional Local Food

Figure 12: Regional production and supply of food

Another crucial aspect of a local sustainable food supply system is to ensure a reduction in the use of artificial fertilisers. Intensive farming practices are less efficient in resource terms than low input and organic techniques. A study by ADAS for the UK government shows that on average there was a 42% energy saving in growing organically.[6] There is great scope for supporting mixed organic farms that produce a good range of produce.

Creating a local food cycle also allows us to return organic wastes—composted green waste and sewage—back to the land, building soils and fertility and supporting organic cultivation. If we don't have a local use for our wastes, we have a waste problem. Instead we could convert a problem into a solution and create a closed loop cycle. It will take some effort to develop efficient and safe ways of returning organic wastes to farms on a more widespread basis. For example, sewage can be contaminated, but sewerage companies are tightening up on what is fed into the sewer system. Another problem is pathogens in sewage, but this can be addressed by methods such as composting, or using human waste solely for fodder crops, pasture or indeed non-food crops like fibre and biomass energy crops.

A good example of closing the loop on the food cycle is demonstrated by a scheme run by Wyecycle in Kent. The company collects organic wastes from homes and these are taken for composting on local farms. Wyecycle also delivers local fresh food and vegetables to people's homes via a box scheme.

Localising the food chain can also reduce other risks such as the spread of diseases. A recent and hugely damaging example in the UK has been the spread of Foot and Mouth Disease (FMD). FMD entered the country via meat from Asia and then spread very rapidly, costing the UK economy somewhere between £ 5–10 billion—as much as the UK earns in 14 years of meat and dairy exports.[7]

Taking into account the costs and risks associated with trying to police a global food industry should make us favour greater localisation. The previous outbreak of FMD in 1967 was much more easily contained and confined mainly to three counties, as cattle and sheep were generally sold and slaughtered close to where they were reared. Today, livestock are transported greater distances as smaller abattoirs have been closed down. Since the spread of diseases is related to the square of transport distance and livestock today may be transported 10 times as far as in 1960s, diseases are 100 times more difficult to control. Then there are concerns about increasing likelihood of pests establishing themselves because of global warming. If temperatures in the UK continue to rise, pests like the Colorado beetle which can be imported on food products, stand a much greater chance of naturalising themselves here.

Much of the current movement of food is pointless. Dr. Caroline Lucas, a Green Party Member of the European Parliament, reports on the food swap between the UK and the Netherlands in a single year (Table 3).[8]

	UK Imports per annum from the Netherlands	UK exports per annum to the Netherlands
POULTRY	61,400 tonnes	33,400 tonnes
PORK	240,000 tonnes	195,000 tonnes
LAMB	125,000 tonnes	102,000 tonnes
MILK	126 million litres	270 million litres

Table 3: Pointless food swaps between the UK and the Netherlands

In a bioregional world we would see much more of our needs met through a local and regional food supply system based on low input and organic agriculture. The food would be produced on local, efficient, mixed farms, producing a diversity of crops, and where animal manure can be used as fertiliser for crops. The food produced on these farms would be delivered through efficient regional packhouses to RDCs for delivery to supermarkets, or for direct supply to consumers. Composted food waste and sewage could be returned to these same farms to create a closed loop cycle.

Chapter 7

Hemp, Clothes and Fair Trade

Textiles are an essential part of our everyday life. We all use clothes, curtains and bedding, mainly made with cotton. Yet most people have no idea that cotton is the most damaging crop grown on our planet. Intensive cotton production is responsible for much human misery and exploitation, and some major environmental problems. Cotton is cheap, but it is the environment and workers in developing countries that are paying the true price. It could all be so different if only we, as a global society, applied some creative thinking, a sense of justice and better land management. However, cotton yields are likely to halve if production is switched to de-intensified, organic and sustainable farming methods,[1] so we need to consider alternative ways of meeting our textile needs in a sustainable future. If we take the BioRegional approach, in the UK and around the world, we could produce more of our textiles locally from low-input fibre crops. In the UK, for instance, we could cultivate our traditional fibre crops—flax and hemp—as we explore later in this chapter.

Textiles used globally include non-renewable synthetic fibres produced from fossil oil such as polyester; renewable synthetics made from wood pulp such as rayon, viscose and tencel; and natural fibres including cotton, wool, flax and hemp. The most popular textile fibre is cotton, which accounts for 33% of global textile production. Fossil oil-based synthetics, as a non-renewable product, are not seen as environmentally friendly. However, a life cycle assessment by clothing company Patagonia found that cotton causes the greatest environmental damage of all textiles.[2]

The trouble arises because although cotton is a very desirable textile fibre, it is a difficult crop to grow. Cultivated on just 2.4% of the world's agricultural land, it accounts for 14% of the world's use of pesticides and 7.5% of the world's use of artificial fertilisers.[3] Cotton will only grow in warm, humid climates, or in warm climates with considerable irrigation: 73% of cotton is produced in irrigated fields, and when calculated per kilogramme of product, cotton is the world's most water-intensive crop.[4] Cotton used in the UK is imported from countries all over the world, including the southern USA, Uzbekistan, Australia, Tanzania, Pakistan,

Mali and Israel.[5] In developing countries, large amounts of precious water and fertile land are used to grow cotton as a cash crop to repay debts, whilst there are shortages of agricultural land and water for food crops to feed the people who live there. For example, in Ethiopia, 60% of the fertile Aswan river valley has been devoted to cotton production. Local people have been forced on to fragile uplands contributing to the deforestation that has been partly responsible for Ethiopia's ecological crisis.[6]

To give another example of damage associated with cotton production: Uzbekistan is the fifth largest producer of cotton, growing 5% of world supply. It was the second largest supplier of cotton to the UK during the 1990s. The Aral Sea in Uzbekistan, once the fourth largest lake in the world, is now only the eighth largest and an ecological disaster area. Between 1960 and 2000 it lost 80% of its volume[7] as its feeder rivers, the Amur Dar'ya and the Syr Dar'ya, were diverted to irrigate the cotton and rice fields. All commercial fishing in the Aral Sea has been lost. Only four species of fish remain where there used to be 24, and just 38 mammal species where there used to be 173.[8] There is a shortage of fresh food because local farmland has deteriorated due to salination and pesticide poisoning. The loss of the cooling effect of the Aral has caused the area to heat up. The health of the local population has suffered with epidemics, soaring cancer rates, increased liver and kidney disease, high rates of infant mortality and birth defects. Dr Leonid Elpiner, who specialises in health problems in the Aral region, characterised the afflictions being experienced in the area as "pesticide AIDS".[9]

When we first started researching this area in 1992, the problems in the Aral region were apparent and the international community established various research and aid projects.[10] We were shocked to discover that 10 years later, the problems have become worse and that Uzbekistan continues to grow and export the same amount of cotton, using the same discredited irrigation and pest control methods.

The problems with intensive cotton production are not confined to one or two regions of the world. In the USA, a survey by the California department of health services found health problems in residents of cotton growing agricultural communities, connected with pesticide and defoliant use.[11] Contaminated run-off from cotton fields in Alabama, USA caused the death of 240,000 fish in 1995 even though the pesticides had been correctly applied. China's Yellow River Valley has falling groundwater[12] and farmers in Xinjiang lost one million hectares of their crop to pests after pesticides killed the beneficial predatory insects.[13] Cotton production nearly destroyed Lake Hula in Israel.

The ecological damage described above is quite damning in itself, but we must also consider the global transportation of textiles and how this trade impacts on less developed countries. A newspaper feature article re-traced the production of a pair of Lee Cooper cotton jeans on sale for £19.95 in a UK store.[14] Journalists found that the materials used had made a journey of 40,000 miles clocking up CO_2 emissions and allowing almost no accountability. The cotton was grown in West Africa and Pakistan, dyed in Italy and sewn in Tunisia. Brass metal for the buttons and rivets was made in Germany from zinc and copper from Australia and Namibia. The zip teeth and thread were from Japan.

The cotton for the Lee Cooper jeans was produced in Benin, one of the main cotton-growing countries in West Africa. During the 2000 cotton season one hundred people died, poisoned by the pesticide endosulfin which is used on the crop. The chemical is so dangerous that it is banned in many countries. In Benin, cotton is grown near to food crops and so poisoning is almost inevitable. The financial reward for local people is low. Cotton pickers earn 60p a day and corrupt government officials sometimes keep small farmer's payments for themselves. Farmers also have to cope with poor yields when the rains fail: one farmer growing cotton on his three-hectare plot made just £15 profit. The farmer's 80-year old father has grown cotton in the area all his life, and says that the only way to make money is to have plenty of family members working for free. The UN Children's Fund states that cotton growing areas in Benin have the highest school drop-out rates. Another newspaper report from 2001 tells how Mr. Bapabiozo, a desperately poor farmer in Benin, sold his four sons, aged from eight to twelve years old, into slavery for just £10 each.[15] The boys were destined for cotton plantations in Ghana, but were saved by charity workers from UNICEF. According to UNICEF, every year 200,000 children from West Africa are sold to work 12-hour days on the cotton or cocoa plantations. This is enough to put anyone off cotton clothes and chocolate bars which are not fair traded.

Cotton is a commodity cash crop, the price of which, like all commodities, fluctuates and has been driven down over time, making it difficult for poorer countries to benefit from their exports. In 1986, the then President of Tanzania, Julius Nyerere said: "This year the rains were quite good . . . the cotton crop has doubled. We are desperately short of foreign exchange and cotton is one of our major exports. . . . But the price of cotton on the world market halved from 68 cents to 34 cents a pound in one day in July this year. Our nation has made the extra effort, but the country is not earning any extra foreign exchange. That is theft."[16] The price of cotton on the world market today, 16 years later, is still just 42 cents per pound.[17]

The inequity of the international commodity trade is further compounded by trade barriers. The UK Department for International Development has acknowledged that trade barriers set up by the EU, USA and World Trade Organisation prevent developing countries from prospering from their hard work. The Secretary of State, Clare Short, noted: "Especially in the areas of agriculture, trade and clothing, developing countries face high tariff and non-tariff barriers. From recent research, developing countries stand to gain $150 billion—three times what they get in aid—if tariffs are cut by half. The UK government therefore strongly supports duty-free access for less developed countries".[18] An Oxfam campaign states that the World Trade Organisation rules are loaded against poor nations. While world trade flows have trebled in the last two decades, the world's 48 poorest countries saw their share of global exports fall by half to just 0.4%.[19] Global trade is currently exploiting many vulnerable countries. The links in the supply chains are so convoluted and the situation so complex, that this exploitation is not a conscious decision by most people involved in the trade. However, this ignorance cannot be an excuse for companies not to make sure that they are supplying products to consumers free of exploitation. Nor should it be an excuse for governments not to ensure that decent social and environmental standards are gradually introduced internationally to start to foster equity and create a level playing field for trade. Fair trade products should be standard, not an option for consumers.

To start addressing these problems, we must work in partnership with developing countries to create a balanced form of economic development. We can help these countries to sell organically produced Fair Trade cotton fabric or garments, tariff-free, to developed countries instead of forcing them to sell raw cotton. A tonne of raw cotton sells for $120, but a tonne of cotton would produce enough fabric to make 120 pairs of jeans worth $840, switching from a low FEET product to a high FEET one (chapter 5). This approach would allow cotton growing regions in developing countries to start freeing themselves from the tyranny of low commodity prices. Given fair and higher incomes from exporting a value-added product, parents could send their children to school instead of working in the fields (education being the key to escape from poverty), enjoy the luxury of family planning, and give time to the consideration of issues like protecting the environment. At the same time, we must support these countries to diversify away from simply exporting cotton, enabling their farmers to produce a greater diversity of fresh food for local consumption instead.

Switching to Fair Trade organic cotton will not be the whole answer as cotton is an exceptionally demanding crop and yields are likely to halve if sustainable and organic farming methods are introduced.[20] We will need to encourage production of other natural fibre textile crops such as flax and hemp, both in developing countries and in countries such as the UK.

In 1994 we began to investigate how we could produce our own BioRegional textiles in the UK.[21] Flax (from which linen is made) and hemp are multi-use annual plants that have been grown for thousands of years to make cloth, paper, rope and oils. We suggested that flax and hemp could be grown on land being taken out of food production and that these fibre crops could be part of a sustainable industrial ecology. The better quality fibres could be used for textile production, the shorter waste fibres for paper pulp and the woody core for animal bedding or particleboard.

The environmental benefits of textile, paper and oil production from hemp have been tested in academic research. A study carried out by the University of Melbourne found that hemp production for textiles, oilseed and paper as an alternative to cotton textiles, oil and forestry would increase economic efficiency whilst reducing the ecological footprint of production of these goods by up to 50%.[22]

Hemp is *Cannabis sativa*, the leaves and flowers of which are a well-known drug. Consequently hemp cultivation is illegal in many countries and had almost ceased in the western world. But plant breeders have developed low narcotic cannabis varieties. During the 1990s an increasing number of countries have permitted low narcotic hemp production. In the UK around 2000 hectares have been grown every year since 1993 to produce animal bedding, cigarette paper pulp and fibre or recyclable car interiors.

Harvesting and processing of hemp and flax fibres is similar, as they are both bast fibres: that is, the fibre is found along the full length of the stem. Unlike hemp, flax production has never ceased in France and Belgium, where they use old-fashioned, inefficient and somewhat expensive harvesting and production methods such as pulling, retting, scutching and wet spinning. Similar methods can be used for hemp, but as the industry had all but died out until recently, the only machines are made in eastern Europe and do not meet EU health and safety standards. China is a large hemp textile producer, but the harvest and fibre extraction are carried out by hand before going for machine spinning and weaving. This would not be economically viable in the EU.

Fabrics made from hemp and flax are very comfortable to wear: they 'breathe' and so keep the wearer cool in hot weather. They are also soft and yet hardwearing: one of the authors has owned a pair of hemp jeans for 10 years. They have not worn out, merely faded, and so have had to be re-dyed three times. The durability of hemp led Levi Strauss to make his first jeans from hemp cloth imported from Nimes in France—hence the name "denim", from "Serge de Nimes".

Flax produces very fine fibres and is a much lower input crop than cotton. But because hemp can be grown organically more easily than flax, and because at two tonnes per hectare it can produce double the quantity of textile fibre, we have concentrated our efforts on hemp.

During 1994-6 we carried out a practical trial to see if we could produce hemp textiles in the UK.[23] A Kent farmer, Chris Older, had already run flax-growing trials on his farm and allowed us to grow hemp on his land. We tested four varieties under the supervision of Wye College, University of London and found that Hungarian varieties had a higher yield than French ones. We saw hemp's remarkable weed smothering properties due to its rapid rate of growth, and the crop did not suffer from any pests. These properties make hemp ideally suited to organic cultivation, adding positively as a clearing crop in an organic rotation, and indeed most hemp in the UK is grown to the low input 'Conservation Grade' standard. We tested the traditional method of processing hemp for textiles and a new more cost competitive method developed for extracting high quantities of fibre from linseed flax straw. However, we obtained the best textile quality fibre from the traditional processing method. The resulting hemp fabric was used by the London designer Katharine Hamnett to produce a summer blazer pictured in the centre of this book.

The conclusion of our trials is that hemp really does flourish in the UK and a quality local organic textile fabric can be produced. However, the methods of harvesting and processing, including the practice of dew retting where the crop is laid on the ground to ret (rot) to enable the fibres to be extracted are uneconomic, inefficient and subject to the vagaries of the weather. More cost effective and reliable harvesting and fibre extraction methods need to be developed.

An Australian company, Fibrenova, has now developed a promising harvesting and fibre extraction technology that could make hemp competitive with cotton in terms of both quality and price. We are now working with Fibrenova to conduct trials on the equipment in the UK. Flax and hemp textile production could give a much needed boost to UK farmers, rural employment and the UK textile industry.

Chapter 8

The Multiple Benefits of Bioregional Development

We believe that creating stable regional economies can help create a sense of community and security that can alleviate the stresses inherent in a globally competitive world. A sense of community can be supported by fostering a sense of place, through locally distinct neighbourhoods and industries linked to the ecology and heritage of an area. We can illustrate these benefits with examples from a number of our projects.

In terms of human organisation, at the extremes of scale we have individual self-sufficiency at one end and globalisation on the other. Aiming to be self-sufficient as an individual is not easy. Neither is it necessarily efficient in terms of time and energy to grow food, make clothes and generate energy on one's own. To be self-sufficient, we would have to live at very low densities, which means that we couldn't run public transport effectively. Higher density living also has other advantages. It is easier and cheaper to insulate a terrace of houses rather than a single home. Self-sufficiency prevents us from developing the specialisation from which we get craftspeople, scientists, artists and so on.

However, it doesn't follow that big is good and that bigger is better. The biggest cities today, such as Tokyo, Mexico and New York, have populations substantially exceeding 10 million. Yet cities until the 20th century were much smaller. Athens in 5th century BC had a population of only 30–40,000. As cities move from being big to becoming mega-cities, their environmental problems also grow. We have to bring in vast quantities of resources from further and further afield to feed and clothe their citizens, which becomes increasingly costly in transport energy terms. Wastes become problematic, because of the sheer quantity and density at which they are produced. We create problems of pollution from traffic, sewage and refuse disposal.

Mega-cities resulting from indiscriminate economic globalisation create a very unbalanced and inequitable form of development. London and the south-east of England have boomed through the 1990s, drawing in people from the rest of the UK and from overseas, many entering illegally as economic migrants. Inevitably this causes discontent within the local population. This discontent is reflected across Europe: witness the rise of right-wing politicians like Le Pen in France and Pim Fortuyn in the Netherlands. So this inequitable development is not only a problem for developing countries but also for the developed world.

Unfortunately it is generally accepted by most of our politicians that London still needs to grow so that it can retain its position as a global financial centre. London is planning for growth from its current 7.4 million population to 8.1 million by 2016.[1] Yet its transport infrastructure is already overloaded. And what does London's prosperity really mean? More homes are being built around London and the south-east, albeit not fast enough to meet demand, and green space is being lost. Roads are congested. House prices have more than doubled in 5 years so that even people in good jobs like nurses and teachers can't afford to purchase homes. The UK government is having to invest £60 million to provide subsidised housing for these 'key workers' to try to maintain public services. Quality of life for people living in London and the South East is deteriorating. Yet there are other regions in the UK, particularly in the Midlands and the North of England, which have problems of empty and abandoned property. These regions are suffering from a deteriorating quality of life because of contraction of their economies. In some parts of the North it is impossible to give homes away. This north–south divide in the UK is highlighted in a report by the Joseph Rowntree Foundation.[2]

Then there is an urban–rural divide, with rural areas unable to share in the prosperity of London. Farmers' incomes fell to an average of only £5,200 in 2001, about 20% of the average UK salary, and the rate of depression and suicides among farmers has increased.

In the UK we have a north–south divide, an urban-rural divide and a 'two speed' economy of success and failure, both of which cause problems. It is a distorting system which, when it is successful on its own terms by displaying economic prosperity, requires further distortions in the form of subsidies (such as for key worker housing) to make it work— and subsidies are rarely an efficient way to allocate resources or to tackle problems.

What route might development take on a more bioregional planet?

By ensuring that the economy is taking into account all the costs of environmental damage, fossil fuels and other non-renewable resources would become more expensive. Transport, particularly by air, would cost more, so that market forces would create an economy in which goods were moved around much less. This would shift production towards a more local and a smaller scale.

In agriculture, the competitiveness of local farmers would increase, creating a healthy farming ring around the city. The proximity of these farms to the cities would mean that the farmers could use organic wastes and sewage from the city for fertilising their fields, creating a symbiotic relationship between city and farms. Farmers might become more imaginative in what they grow, raising locally what were previously considered exotic or unusual crops—Chinese greens, okra, squashes, new peppers and sweet potatoes. They might extend growing seasons by raising more crops locally under glass, using heat currently wasted at power stations.

In industry, the economy would favour local production of bulk commodities. Local paper recycling would become the norm. Losing the economies of scale by moving to smaller scale production in reality would simply shift employment away from the transport sector to jobs in recycling, local manufacturing, farming and forestry. Whilst smaller scale production might increase labour costs per unit of production, these would be offset by lower investment costs and greater adaptability to local conditions. Gaining planning permission for a small factory is always easier than getting permission for a large-scale one.

Creating a more balanced regional, self-regulating, diverse and stable economy will create greater richness in opportunities for people to chose a wider range of careers and vocations. The connection between quality of life and economic diversity will become increasingly evident.

Other benefits might also accrue from bioregional development. We have seen an increase in disillusionment with politics in the UK. Although there are many factors contributing to low turnouts, loss of influence seems to be contributing to a sense of apathy. As people increasingly believe that corporations, rather than governments, are in control, low election turnouts are perhaps not surprising. Regional scale development encourages people to become engaged, creating an environment in which the political ideal of subsidiarity can be expressed.

Bioregional development also reduces risk. Economic risk increases with specialisation, when we put all our eggs in one large basket. A country or region dependent heavily on one product or service is of course very vulnerable. For example, the valleys in South Wales were badly hit in 2000 by the closure of five clothing factories when Marks and Spencer decided to source their products from less expensive manufacturers in North Africa and Eastern Europe. The GMB union said the switch to cheaper suppliers was "designed to please the City of London at the expense of Britain's clothing industry".[3] In another example, the slump in cotton prices hit Tanzania very badly in 1987, as we described in chapter 7. Transnational companies risk global scale mismanagement, as was evident in the collapse of energy giant Enron in 2001. The social and financial costs and risks are huge, and have to be taken into account when assessing objectively our current development paradigm. Global markets inevitably suffer from global swings, creating local instability. But society has much to gain from a sense of continuity and stability that allows each generation to build on the work of the previous one. Whilst opportunists will always profit from boom and bust development, it is expensive in human terms.

In contrast, a diverse regional economy is more stable. The positive effect of any injection of money into an economy is multiplied as the money is spent and re-spent locally. The level of this positive effect, known to economists as the local multiplier, is dependent on the number of times the money is re-spent before it leaves the community. The local multiplier is therefore larger in a diverse local economy that offers a range of products and services. We can illustrate this using the example of some BioRegional products and services, and showing how they might contribute to a locally sustaining economy. Local recycled paper is purchased from the mill which employs 300 people in Kent. A further 200 people are employed in collecting the waste paper. These workers can in turn spend their income on other local products such as local Kentish strawberries and charcoal, supporting these businesses and in turn creating a demand for local paper—a self-supporting economy. In the global forests-to-bin scenario illustrated in Chapter 3, we have exported the jobs and the money. An Asian multinational company profits from the paper sales. The only money which stays in the local economy is from jobs in the refuse collection service, landfill management and waste to energy incinerators. As we described in Chapter 3, recycling paper generates ten times the income for the UK and creates three times as many jobs as energy generation from waste.

To take another example, if we build zero energy homes using reclaimed and local materials as we have done profitably at BedZED, we will create jobs and income for local reclaimed materials yards, brick makers and in woodlands. The alternative is overflowing landfill sites, derelict woodlands and more holes in the ground from quarrying.

As well as economic risk, globalisation results in increased environmental and health risk. Global warming is one clear example of environmental risk. Another area of risk is in the spread of disease. In chapter 6, we described the huge costs of Foot and Mouth Disease to the UK and the fact that globalisation and transport were at the root of the epidemic. The problem of global spread of diseases and pests is not limited to farming. It is a problem in forestry—examples include the introduction of the European gypsy moth in the USA. Human diseases are also spreading much more rapidly. The UK government's Chief Medical Officer, Sir Liam Donaldson, recommends setting up an expert body to warn the public about the increasing risks of tropical diseases caused by increased travel and climate change—diseases like Ebola virus, Dengue fever, West Nile virus, tuberculosis and malaria.[4] Professor Donaldson also warns: "Given the nature of micro-organisms that cause infection, the path of human behaviour and changes to the environment, further newly emergent diseases are inevitable. It is essential to expect the unexpected." With no duty on aviation fuel and cheap plane tickets, we are creating a way of living which involves regular air travel. In doing so we are becoming a society dependent on air travel and multiplying our vulnerability to disease.

Although harder to quantify than the costs of the spread of diseases, there are real health, and hence economic, benefits to a sense of well-being. Health is not simply the absence of disease. Local economic development creates a sense of community and place, which adds to a sense of well-being and can be an antidote to the stresses of modern day living. We can enhance a sense of place by fostering its connection to the local ecology, natural resources or heritage of an area. Obvious examples are areas of natural beauty which have become tourist destinations. However, there are opportunities for creating a sense of place on a smaller neighbourhood scale. For example we have been surprised by the effect of a BioRegional project to revive London's historic lavender industry.

In the eighteenth and nineteenth century, the area comprising Mitcham, Carshalton and Wallington in south London was internationally renowned for its lavender industry. Companies like Potter & Moore and Yardley of London built their reputation on the area's high quality lavender oil. Even French companies bought their highest quality lavender oil from the area. However, with the spread of London's rail network, the industry fell into decline as dormitory suburbs were built on the lavender fields.

BioRegional started to work to revive the industry. In 1994 we put out press releases in the local newspapers to see if people had lavender bushes in their gardens which they believed came from the original fields. A number of local residents responded and volunteers collected cuttings from their plants. We formed a partnership with the horticultural unit at the local prison, HMP Downview, with the inmates raising the cuttings. As part of a rehabilitation programme, the inmates were allowed out of the prison on day release to help plant up three acres of disused allotments, offered free of charge by Sutton Council.

We harvested our first crop of lavender in 1998. The event generated a lot of media coverage. Older residents sent us their reminiscences. The new Christmas lights in Wallington have been designed on a lavender theme. When a local pub was refurbished, it chose the lavender story, naming the pub 'The John Jakson' after a family of lavender growers; and when a Sainsbury's supermarket was built locally, lavender was chosen as the subject of a large sculpture at the entrance. We have also worked with Cranfield University to build a novel small-scale lavender harvester.

Yardley of London learned of our project and started sponsoring our 'pick your own' lavender days in 2000. When we distilled our first oil, we also started supplying a limited edition Yardley/BioRegional essential oil. This partnership has grown and we are now working with Yardley to establish a larger 20-acre lavender farm in the area. Homes backing on to our lavender field are now being sold with this as a selling feature, adding value to the homes in the area.

The lavender project demonstrates a number of principles. By utilising disused urban land productively, we are protecting it from possible development. There has been a marked reduction in fly tipping and vandalism on the site as we have brought the site out of neglect. We are showing how we can gain an economic value out of a traditional low-input local crop well suited to a particular soil and climate. We have been able to use heritage as a way of engaging people in a broader environmental issue. The ecology and heritage combination has enabled us to create a project bringing together partners as diverse as a local prison and an international fragrance company. The environmental gains are not huge compared to our Local Paper for London scheme, but the project demonstrates other wider factors which come into play to support a project developed along bioregional lines. The more connections we make between people, the past, the future and the environment, the greater the sense of meaning it brings to people and the greater the chance of a project becoming successful.

Another of our projects, Urban Forestry, has also been building connections with the local community. It has pioneered the involvement of a voluntary conservation group in monitoring the biodiversity of woodlands which in turn are managed for firewood and charcoal supply to B&Q. Rather than just monitoring for records, the work of the Croydon branch of the British Trust for Conservation Volunteers forms an integral part of achieving Forest Stewardship Council (FSC) standards for

Croydon Council's woods. Without this voluntary monitoring work, the additional costs would make many of Croydon's small woodlands uneconomic to include in the scheme. This in turn would make it impossible to supply to B&Q, a company requiring environmental certification for all its wood suppliers.

In a world first, our Urban Forestry project has also extended FSC certification to street and park trees, making Croydon the only urban area worldwide formally classified as a sustainably managed forest. Working with Croydon, we are now converting around 1,000 tonnes per year of tree surgery waste into woodchip to fuel the combined heat and power plant at our BedZED eco-village, generating 130 kW of electrical and 250 kW of thermal energy. London as a whole could recover 100,000 tonnes of tree surgery waste every year for energy generation.[5]

We can envisage a future where we manage street trees along forestry lines, using the main trunk for high value timber and the branches and twigs for woodchip. In our Urban Forestry project we have experimented with introducing a mobile sawmill for processing quality logs. Our practical experience suggests that this could be economically viable if sufficient volumes of wood can be sawn and local markets developed. In the right locations, there are also opportunities to further increase the productivity of our neighbourhoods by planting fruit trees in our streets and parks for use by people and wildlife.

Promoting local and regional scale development can bring many benefits. Some of these benefits are quantifiable or potentially quantifiable—such as reducing our contribution to global warming and various aspects of risk avoidance. Some benefits, however, are not easily quantifiable if at all, such as the effects on health and well-being of creating a sense of place and belonging. However, concepts like quality of life are not simply woolly ideas. Just because we can't quantify something, it doesn't mean it doesn't have economic consequences.

Chapter 9

Bioregional Economics: Comparative Advantage versus BioRegional Advantage

Our society is built on an economic ideology based on exploiting economies of scale, international competition and comparative advantage. These have led us to create a global market which has its benefits and drawbacks. However, the price of many of the products and services we buy does not take into account the damage they cause to the environment, to people and communities. If we were to take these external costs into account, we would see the balance shifting towards smaller scale, more diverse local and regional development, or bioregional development. We can then reap the benefits of bioregional advantage.

It will not take much to tip the balance and make bioregional development a much bigger part of the mainstream economy. Simply making landfill or transport more expensive would favour more sustainable regional development. As we saw in chapter 3, as soon as landfill tax rises to £35 per tonne from its current level of £13 per tonne, local recycling and buy-back schemes such as our Local Paper for London and Local Paper for Scotland schemes become the cheapest office paper supply option. This level of landfill tax is common in many other European countries. In another example, the introduction of a UK Aggregates Tax in April 2002 of only £1.60 per tonne has meant that Edenway Ltd, construction contractors, are now recommending using recycled aggregate and recycled crushed glass sand on all their other projects, having had experience of using the products through their work at BedZED.

In the case of charcoal, transport costs make up 8% of the wholesale price of a bag of local charcoal but 70% of the price of imported South African charcoal. If we double transport costs by introducing further carbon taxes or road tolls, the wholesale prices of a 3 kg bag of imported charcoal will rise by £1.19, and the local charcoal by only £0.22. Their respective wholesale prices now become £2.96 and £2.90. Doubling transport costs makes local charcoal competitive purely on price. Such an increase in transport costs is not out of the question if we are to reduce CO_2 emissions by the suggested Royal Commission target of 60% by 2050 (see Chapter 2).

Of course, increasing transport costs and only trading fairly with developing countries will impact more heavily upon the poorer members of our society in the UK because the price of goods will increase. However, if we are seeing the effect of the local multiplier of money spent and re-spent locally in a diverse and productive bioregional economy (Chapter 8) then we would expect to see an increase in the health of the regional economy, good levels of employment and stable communities where we could choose to provide increased support for more vulnerable members of society. From a broader perspective, it cannot be moral for us to use the excuse of supporting our own poor by exploiting even poorer people in other countries. Someone somewhere pays the price for cheap goods. Furthermore, if we use the need to supply cheap goods at the cost of the environment, it is the poor who will suffer most in the medium term. The poor throughout the world are those least able to buy their way out of environmental problems as they arise—whether it be people flooded out of homes in Lewes, East Sussex, England, or from homes in Bangladesh. In the long term all of us, rich and poor, will suffer.

Although we support market mechanisms as the way to deliver products and services, we cannot leave the market to deliver everything we need. We all need a healthy planet if we are live peaceably on it. But important environmental benefits cannot be bought by an individual as a consumer. Clean air and a stable climate are public services—like streetlighting and the police which have to be paid for by society as a whole. Most consumers quite rightly are not willing to pay more for goods or services, like green electricity, which benefit everyone. It is therefore wholly proper, even within classical market economics, to introduce regulation (like placing limits to CO_2 emissions) and to deploy taxes like road tolls and carbon taxes to correct areas of 'market failure'.

Deploying regulation and taxation to deliver a healthy planet does not mean that economic collapse and unemployment will automatically follow. As we suggest in chapter 8, in many cases we will simply replace international trade in bulky commodity products with local production and local recycling. There may then be fewer jobs in the transport sector but these will be compensated for by an increase in local farming, forestry and manufacturing jobs. We may need to be less wasteful, but any loss in material wealth may be replaced by an increase in quality of life. There will be fewer lorries on the roads and greater opportunities for communities to develop around new more regionally based economies. With some protection from global competition, local markets for commodity products will be more stable, giving farmers and foresters a fair chance to take a long term view of developing their industries. High value products like computers and medicines (high FEET products), whose transport costs make up only a small percentage of the their price, will hardly be affected by increases in transport costs. Markets for these products will remain global and therefore the scope for innovation and product development will not be reduced. We can remain attached to economic globalisation at all costs or, alternatively, we can start behaving rationally, creating a healthy balance between global trade in high value products with local trade in everyday goods.

Much of the anger in the green movement is directed against economic globalisation. Edward Goldsmith,[1] George Monbiot[2] and Anita Roddick[3] have documented some of problems caused by indiscriminate globalisation, which they blame largely on the behaviour of transnational corporations. Others like billionaire financier George Soros[4] and Adair Turner,[5] Director General of the UK Confederation of British Industry, recognise the grievances of the anti-globalisation movement, but attribute the main problem to a failure in international governance. They argue for a strengthening of international governmental institutions to regulate the market in which transnational corporations operate so that it makes sustainable development possible.

Sustainability certainly won't be achieved on the basis of current economic development models. In his foreword to London's draft Spatial Development Strategy (SDS), mayor Ken Livingstone says,

> I hope that everyone with an interest in securing London's future success will take part in the development of the London Plan and work together to ensure London becomes the exemplary sustainable world city that I would like it to be.

However, the draft SDS is promoting expansion of all five London airports, so that the city can maintain its position as a global financial centre and "ensuring that expanded airport capacity brings economic, social and environmental benefits". It is very hard to see how expanding airport capacity can improve London's environment and make London an exemplary sustainable city.

When it is so clear to us that an indiscriminate attachment to globalisation is misguided, why does it form such a core component of current economic and political thought? At the heart of the ideology is the concept of comparative advantage, which underpins international trade. Of course, we can't easily or efficiently grow coconuts in the UK so if we want to eat them we need to import them from the tropics. We trade in specialist goods like Chinese silk, French wine, Namibian uranium, Scottish salmon, Russian caviar and Saudi oil. A particular natural resource or a favourable climate might confer an absolute advantage to some countries. Comparative advantage, however, is a concept that goes further than absolute advantage.

Comparative advantage states that a nation gains by doing what it can do best relative to other nations. The concept was developed by the economist David Ricardo in the early 19th century. He proposed the hypothetical model of cloth and wine production in a two-country world comprising England and Portugal.[6] In his model he set the man-hours required to produce each unit output of cloth and wine as follows:

Man hours required per unit of output

	Portugal	England
Cloth	90	100
Wine	80	120

To produce each unit of cloth in Portugal takes 90 man-hours while in England it takes 100 man-hours. Portugal is even more relatively efficient in the case of wine production. Each unit of wine takes 80 man-hours in Portugal and 120 man-hours in England. In Ricardo's model, Portugal has an absolute advantage in both cloth and wine production.

At first sight we may assume that Portugal wouldn't gain from trade with England. But what happens if the two countries exchange a unit of English cloth for a unit of Portuguese wine? In line with our intuition, England gains by saving 20 man-hours, as it takes only 100 man-hours to produce the cloth in England, but 120 man-hours to produce the wine. But counter-intuitively Portugal also gains. When Portugal imports one unit of English cloth, it frees up 90 man-hours of work, i.e. the time it takes to produce one unit of cloth in Portugal. It takes only 80 of these man-hours to produce the unit of wine required for the exchange. So Portugal has saved 10 man-hours by making the exchange. Although Portugal has an absolute advantage in both cloth and wine, Portugal does gain, in Ricardo's model from trade with England.

Ricardo's model is simple and powerful. However, its simplicity turns out to be a problem for three reasons.

First, Ricardo's model assumes that capital does not flow between countries (which it didn't to any great extent in Ricardo's world of the nineteenth century). Therefore every country would retain some industry. Today however capital flows very rapidly across the world, in the words of international financier George Soros, like a "wrecking ball".[7] Individual nation states have to vie with each other to create an economic climate which will attract the capital. In order to do this there is pressure to reduce labour and production costs, causing social and environmental problems.

Second, Ricardo himself clearly states that benefits of specialisation and comparative advantage apply only in so far as there are no costs to transport. There are of course transport costs to international trade—financial, environmental, health and social ones. Burning fossil fuels for transport is the fastest growing contributor to global warming. Ironically, Ricardo's two countries (England and Portugal) are both suffering the costs of global warming. There has been severe flood and storm damage over the past two years in England linked to global warming. It has cost heavily in terms of money, man-hours and human misery, making some homes uninsurable and hence unmortgageable. Similarly, global warming has exacerbated droughts in Portugal and even more ironically reduced grape harvests in some areas and the comparative advantage of the Portuguese wine industry. Then there is a very wide range of health costs—from cancer to stress—associated with our increasing globalised economy.[8]

The third problem with comparative advantage is that it undermines our scope to use resources efficiently. International trade makes it harder to build industrial ecology and create local nutrient and energy cycles. To build industrial ecology we need to create links between industries and between industries and consumers. This requires diversity within an economy which specialisation does not allow.

We can illustrate this last point by expanding Ricardo's model. Grapes are crushed to make wine, leaving skins as a waste product. In our expanded model the wine industries in England and Portugal use the waste grape skins to make compost. In both countries, the compost is used in local market gardens to grow fresh vegetables to supply the local market. Let us also assume that the cloth in both England and Portugal is made from sheep's wool. As well as providing wool, the sheep also provide local people with fresh meat. There is a high level of recycling and re-use—in both countries the local community return their wine bottles back to the wineries where they can be refilled. Both England and Portugal have diverse local industries—inter-linked and organised sustainably on the basis of industrial ecology, recycling and using wastes as resources. We describe these economies as exhibiting BioRegional Advantage.

What happens if we exploit the apparent advantages of Portugal specialising in wine and England in cloth? We close down the wineries in England and the textile mills in Portugal. With the textile mills closed in Portugal, most of the clothworkers go to work in the wineries. Some become lorry drivers, taking bottles of wine to England and bringing back cloth with them. However, the Portuguese sheep farmers are at a loss—they can't sell their wool and the local economy is in danger of losing a local source of meat if the industry collapses.

In England the wine makers have been retrained in clothworking. The markets gardeners, however, have lost their old source of compost made from the waste grapeskins. The English economy is losing a source of local, fresh vegetables.

In the short term we have a profitable Portuguese wine industry and a profitable English cloth industry. But we also have alienated Portuguese sheep farmers and created desperate English market gardeners. How might this story unfold?

The Portuguese government, fearing loss of the sheep farmers' vote, gives them a grant to establish a new wool marketing board. The wool marketing board engages an expensive business consultant who identifies England with its flourishing cloth industry as a potential market for wool. However, to afford the transport costs to take the wool to England, the Portuguese sheep farmers will need further annual subsidies. The Portuguese government levies a duty on wine to raise the funds to pay the sheep farmers. Now the Portuguese wine makers are unhappy, having to support "whingeing sheep farmers who can't adapt to change and progress". Exporting wine has meant Portugal needs to export wool as well. Portugal has become dependent on wine exports, wool exports, taxes and subsidies.

The vast increase in the number of lorries is also causing trouble on the roads which weren't built for heavy traffic. A new government is elected, promising to upgrade the roads and introduces a super tax on higher earners to pay for the improvements. The business consultant is upset about having to pay the new tax. Meanwhile, back at the Portuguese wineries there is a growing mountain of waste grapeskins, generated at a faster rate than the market gardeners can compost them. There is already a vegetable surplus locally which has forced prices down. To maintain profitability the market gardeners are thinking about exporting vegetables to England. But road freight to England is too slow for the vegetables to arrive fresh—so the Portuguese are considering air freighting them. The business consultant seizes the opportunity. He coins the term 'industrial ecology' for a new waste management tool which could deal with the grapeskin waste problem. The winery bosses secure a grant from the government to engage the business consultant to identify possible industries that could use the grapeskin waste, turning 'this problem into a solution' and creating jobs in new industries for the struggling sheep farmers.

In England, things are not much better. With the loss of their grapeskin waste, the market gardeners have no raw material from which to make compost. Instead, a large chemical company sets up a plant producing artificial fertiliser to sell to the market gardeners. New taxes have been introduced, because like Portugal, England has to invest in new road infrastructure. With the economy so dependent on oil to fuel its lorries, the English establish a military base in Gibraltar to protect their interests in the oil fields of North Africa and the Middle East. Back home in England, the local Friends of the Earth group are frustrated at the public's poor response to its wine bottle re-use and recycling campaign. The local authority responsible for waste disposal say there is little point in collecting used bottles as it is uneconomic to transport them to Portugal where they could be refilled or recycled. Friends of the Earth try to point out that at the same time as a new landfill is being dug locally to bury the waste glass, a previously unspoilt area in Portugal is being quarried for silica for glass to produce new wine bottles for use in England. A government scientist links the drop in fish numbers in the local river to effluent from the fertiliser plant. Another scientist links the increased incidence in droughts which are reducing grape harvests to the increase in atmospheric CO_2 caused by air-freighting fresh vegetables to the UK.

In our revised model we see that we have exchanged sanity and bioregional advantage for comparative advantage and indiscriminate globalisation. Although our revised model describes a worst case scenario, these lessons translate clearly into real life where all around us we can see can the breakdown of local nutrient and energy cycles and the social and environmental problems that result. We don't re-use bottles and have a waste glass mountain. Green waste is not composted and used for food growing, so waste is a problem and has to be sent to land-fill. Where we don't have a local paper recycling mill but have to transport waste paper great distances, it is not economic to operate a waste paper collection. Because we don't have direct local links between farms and consumers we can't re-use packaging such as egg cartons, but have to throw them away. If we have one country specialising in textile manufacture and another in pulping paper, we reduce the opportunity to use textile waste as raw material for paper-making. Specialisation and comparative advantage breaks links in cycles. It does the opposite of industrial ecology—by taking away the links between industries it converts resources into wastes.

Economist J.M. Keynes had this to say:

> I sympathise, therefore, with those who would minimise, rather than with those who would maximise, economic entanglement between nations. Ideas, knowledge, art, hospitality, travel—these are the things which should of their nature be international. But let goods be home-spun whenever it is reasonably and conveniently possible; and, above all, let finance be primarily national.

Arguing for a return to more local production is not arguing against progress. If combined with a new understanding of its value and more efficient technologies, it is progress. If Ricardo were alive today and knew what we know now, would he be promoting comparative advantage? Probably—but only for a much more limited set of products and services.

Chapter 10

BedZED: Living off One Planet

Fifty per cent of the world's population now live in cities, which currently account for around 75% of all resources consumed and wastes produced.[1] The proportion of people living in cities is forecast to grow to 60–70% in this century. Therefore, making our cities sustainable is one of our greatest challenges. Beddington Zero Energy Development (BedZED) is demonstrating how we can create high quality urban environments and live within our 1.9 hectare target ecological footprint.

We can draw parallels between cities and living organisms. Cities consume nutrients and produce wastes. Herbert Girardet has calculated the material flows in and out of London and used this information to estimate London's ecological footprint, which he quantifies at around 125 times the land area of London itself.[2] A more detailed ecological footprint for London is now being calculated in the City Limits London project (www.citylimitslondon.com).

Most cities today display a linear metabolism: a one way flow with resources coming in and wastes pumped out. Food is brought into cities, eaten and then sewage is discharged into rivers, coastal waters or is burned. Inputs and outputs are more or less unconnected. To become sustainable, cities need to evolve a circular metabolism where waste is used as a resource rather than discarded as a nuisance. Sewage should be returned to local farmland to grow the next crops of food, or to grow biomass energy crops, for the city. Paper waste needs to be recycled back into paper. Our BioRegional Local Paper for London and Local Paper for Scotland schemes, described in Chapter 3, are examples of how we can bring a circular metabolism into being. This circular flow mimics natural nutrient cycles so that cities can start working with nature and contributing to the health of the planet.

We can draw another analogy, if rather emotive, between the uncontrolled growth of cities with the growth of tumours in the body. A tumour, like a city, needs nutrients to maintain it and fuel its growth so draws blood vessels, or roads, to serve it. The blood vessels get larger as the tumour grows, forming a dense network around the tumour (*rete mirabile*, or 'amazing network'), resembling the congested ring roads around major cities—the M25 motorway around London is one such example. As the tumour continues to grow uncontrolled, it physically cannot draw in sufficient nutrients to support its metabolism, nor remove wastes from the centre. The tumour dies in places at the centre—so-called central necrosis. There are some parallels to the inner city decay that has afflicted poorer parts of our big cities. Overcrowding and pollution caused by the accumulation of wastes certainly contribute to health problems in city areas. The health of people living in the UK's cities is worse than those in rural areas. A tumour fails to obey the controls on growth needed to maintain a healthy body. In the same way, our cities are failing to obey the controls needed to maintain a healthy planet. On one hand the capacity of our cities to grow is a great triumph of human ingenuity, science, technology and logistics. On the other hand, we need the wisdom to develop the right conditions to place controls on growth and consumption.

It is, however, possible for us to envisage a sustainable city. Living at high densities can be much more efficient in terms of using resources. It is easier to provide public transport and efficient energy supplies. One of the first principles of planning for sustainable cities is to build at high densities around transport interchanges where it is easier to live without a car. We can also clean up our cities, as many have already done. The air quality in London has improved since coal fires were banned in the 1950s and stricter controls on car emission controls phased in since the 1980s. However, cities consume more than ever and are sending more wastes to landfill.

Building a circular metabolism is one principle we can use to make our cities more sustainable. But we also need to think about how we lay out our cities. The architect Lord Rogers, who led the UK government's Urban Task Force, describes the concept of the Compact City in his book *Cities for a Small Planet*. The Compact City is made up of high-density neighbourhoods linked by efficient public transport systems. The neighbourhoods integrate a variety of residential, commercial and leisure uses, where people can live, work and play. Most facilities are within walking distance, and public transport nodes at the centre of each neighbourhood make it easy to live without a car. Heat and power is generated locally. Rogers links his Compact City to a local countryside creating a circular metabolism for it. He also argues for diversity in the economy of the Compact City, reflecting much of what we promote as bioregional development.

Cities should be about the people they shelter, about face-to-face contact, about condensing the ferment of human activity, about generating
and expressing local culture. . . . Another benefit of compactness is that
the countryside itself is protected from the encroachment of urban
development. . . . The concentration of diverse activities, rather than the
grouping of similar activities, can make more efficient use of energy.[3]

Commissioned by the Mayor of Shanghai, Rogers proposed a Compact
City form for the expansion of Shanghai. His transport engineers calculated that the broader mix of activities would reduce the need for car
use by some 60%. The diverse commercial and residential approach
would also safeguard the city from the boom–bust cycles of the international market. However, Rogers was not given the opportunity to put
the vision and principles of the Compact City into practice. Instead, he
had to take commissions for more unsustainable projects such as
Terminal Five at London's Heathrow Airport. Architects and planners
can't create sustainable cities unless we as a society direct and support
them in that aim.

We have to be careful as to how we plan our futures, or we can get
stuck with highly unsustainable forms of development. For instance,
out-of-town supermarkets have been built in many places which people
can only reach by car. Meanwhile town centre grocery stores are lost.
Children can't walk or cycle to school because the roads are too busy
and dangerous. It becomes increasingly difficult for people to live without a car. Health problems such as obesity and diabetes have become
characteristic of car-dependent societies.

BedZED was created to provide a holistic solution to the problems of
sustainable urban living.

In Chapter 1, we described the energy strategy at BedZED in some
detail. However, super-insulating homes and making them airtight can
create problems in internal air quality. Correcting this potential problem
by ventilating the homes with electrical fans would do no good to the
energy profile of the homes. The design team on BedZED therefore
worked to develop a passive ventilation system which uses wind, even
at very low speed, passing over the homes to exchange stale air for
fresh. The windcowls, which have become a distinctive feature of
BedZED and a sometimes controversial local landmark, point into the
wind and air is forced down into the rooms. Stale air is drawn out from
the kitchen, bathrooms and toilets (removing odours and excess mois-
ture) via the rear of the windcowls, where negative pressure is created

by the wind flow. This negative pressure is accentuated by the curved roofline which has been refined in a wind tunnel, generating lift in the same way as an aeroplane wing, which helps to draw air up out of the homes. A heat exchanger extracts any warmth or coolth from the outgoing air, maintaining the internal temperature of the homes.

To reduce energy consumption in the homes further, we have maximised natural light access into the homes. High natural light levels are promoted by the conservatories, but we have also incorporated skylights which bring light to the back of the homes. The homes are also fitted with low energy appliances (such as A-rated fridges and freezers) and compact fluorescent light bulbs. We anticipate that we will be reducing the electrical demand of the homes by between 30–50%, depending on how people use them.

Unlike homes which have a net heating requirement, offices tend to suffer from overheating. In summer, they often become uncomfortable because of the high levels of body heat and waste heat from computers and photocopiers. Office cooling and air conditioning is very energy-intensive, and accounts for an increasing proportion of CO_2 emissions. To prevent overheating, offices at BedZED have been placed in the shade zone of the homes—creating a mixed use terrace with homes on the south and offices to the north. The offices are provided with natural light via skylights—which create a diffuse but bright north light, well suited to people are working on computer screens.

These terraces of homes and offices have also allowed us to use the roof surfaces of the offices as 'sky-gardens' for the home: maintaining greenspace and bioproductive land in a high density development. The 30 cm of soil on the office roofs also doubles as insulation. Almost every flat gets a private garden (unheard of in modern developments at this density). Although it costs more to build roof gardens, we increase the value of the development by selling the flats for the price of a flat with a garden.

Having reduced overall heat and electrical demand of the homes and offices by half, we were in a position to be able to afford renewable energy without residents having to pay higher bills. Our main energy source is a combined heat and power (CHP) plant which takes chipped tree surgery waste from our Urban Forestry project with neighbouring Croydon Council (chapter 8). The woodchip is delivered to a chip store and a 'seeing' robotic arm lifts them into a gasifier where it is converted into wood gas. The wood gas fuels a diesel lorry engine, which in turn generates electricity. The CHP has been sized so that over the course of a year it generates sufficient electricity to provide for all of BedZED's needs, making BedZED a zero fossil energy development. BedZED therefore makes no net contribution to global warming: it is a carbon neutral development. Waste heat from electricity generation provides hot water for the homes and offices and is distributed via insulated pipes across the site forming a district heating system. CHP is some 30% more efficient than conventional electricity generation as we are making productive use of the heat energy that would otherwise be wasted. Generating on-site also avoids the energy losses in transporting electricity via the high voltage national grid.

BedZED is nonetheless connected to the mains electricity supply so that we can export our surplus electricity and buy it back at times when peak demand is greater than our average generation capacity. This also means that we always have a back-up supply so residents won't be without electricity when the plant is down for maintenance or for modification. One of the major advantages arising from BedZED being a mixed-use development, is that electrical and heat demand is much more even over the course of a day than in a purely residential or purely office development. In a purely residential development there is high demand in the early morning and evening, while offices have peak demands during the day. By moving to renewable energy, we are also future-proofing residents against rises in the price of fossil fuel energy and placing them in a position to make major savings as the relative price of renewable energy comes down.

We have worked hard to select construction materials to reduce environmental impact. Almost all the structural steel has been reclaimed from demolition sites, sand-blasted, cut to size and repainted. We estimate that the fossil energy saved by reclaiming 120 tonnes of steel equates to an ecological footprint of 80 hectares of forest. We also tried using reclaimed doors, but couldn't find enough for this scale of development. Instead we used FSC-certified doors from B&Q. All the bulk materials were sourced from within a 35-mile radius to reduce transport and support the local economy: brick came from Cranleigh in Surrey, and we used local oak weatherboarding, some of it from our own Urban Forestry project in Croydon. Reclaimed aggregate was used for the road sub-base and recycled crushed green glass for sand used for bedding paving slabs.

As we explained in the first chapter, we can't build a sustainable future for ourselves on this planet simply by building energy-efficient homes and offices. We have to develop a whole green lifestyle, which is the primary aim behind BedZED.

Reducing car dependence has been a major guiding principle, and we have set a target to make it easy for residents to reduce car use by half. Having offices on the same site as homes means we have created opportunities to live and work on site. BioRegional and Bill Dunster Architects are relocating our offices to BedZED, and some BioRegional staff, including the authors, are buying properties at the development. Living and working on-site eliminates the need to commute and therefore reduces our eco-footprint. In the past in the UK we have segregated residential and commercial areas, forcing people to commute and creating rush hours. Spending two to three hours each day commuting on a train to and from London does not contribute to a high quality of life.

Although we can walk, cycle or use public transport for most of our journeys, for some journeys a car is the only real alternative. Cars provide so-called 'mobility insurance', but once we have bought a car we tend to use it for all journeys. In a joint venture with SmartMoves, BioRegional have set up a car club at BedZED called ZEDcars. The service allows residents to hire a car by the hour when they need one, thereby providing mobility insurance without needing to own a car. This leaves residents free to walk, cycle or use public transport for the majority of their journeys. Hopefully this will also contribute to the health of residents by integrating more exercise into a daily routine, when lack of exercise is contributing to an epidemic of obesity and diabetes in the general population. The car club can also save residents money. The costs of private car ownership are very high—insurance, depreciation, road tax, servicing, repairs and fuel. A person can save up to £1,500 per year by joining the car club rather than owning an equivalent new car. Although car pools are new to the UK, they have been operating for many years on the continent of Europe. A car pool with a range of vehicles from small cars to people carriers, from conventional to electric cars, allows users to choose the vehicle most appropriate for their journey. Some car pools on the Continent also offer benefits like shared access to boats and trailers—greatly expanding the attractiveness of becoming a member.

BedZED is not a car-free development, as could be done in a more central London location with more frequent and flexible public transport. BedZED residents can own a car, but need a permit—not unusual now in developments across London. Instead we are maximising opportunities to avoid using a car. We have organised discounts at local cycle stores to encourage bike ownership. To reduce the need to make a regular car journey to the supermarket, we are negotiating bulk home deliveries of groceries from supermarket chains. This includes preferred delivery slots to BedZED, enabling them to deliver to more than one home at a time, reducing the transport footprint (and saving the supermarket money in the process). LPG-fuelled vehicles, rather than petrol or diesel ones, are being planned to make these deliveries, reducing local air pollution and the ecological footprint even further. BioRegional have also contacted local organic farms and organised direct delivery of local, seasonal vegetable and fruit boxes. Removing a regular car journey—the trip to the supermarket—reduces car dependency, and leaves the car idle, increasing the cost per mile to run it. We play a game of probabilities, all the time finding ways to tip favour away from cars, without banning cars outright. Over the past 50 years the opposite trend has prevailed, so that for many people it is impossible to envisage life without the car. We hope BedZED will demonstrate that if we want to, we can turn things around, but we need a coherent and comprehensive strategy.

We are also promoting the use of electric vehicles at BedZED. We have fitted $777m^2$ of BP Solar photovoltaic panels connected to electric car charging points. The solar panels are sandwiched within the double glazing in the top floor conservatories, generating electricity at the same time as providing shade for the conservatories, preventing the conservatories overheating in the height of the summer. We will generate enough solar electricity to run up to 40 electric vehicles, each providing 8,500 km per year of carbon neutral motoring.

Reducing car dependence at BedZED has allowed us to negotiate with the local council to reduce the number of car parking places from the usual 160 parking spaces for a development of this size to 84. In return we are legally bound to provide a car club. Reducing the number of parking spaces has meant we have been able to build additional housing instead—a developer's dream to increase financial returns on an urban site. This forms another part of the financial proposition which makes BedZED pay its way.

BedZED also addresses water efficiency. Rainwater is harvested from the gardens and roof surfaces and stored in tanks in the foundations of the terraces. This water is used for flushing toilets and irrigating gardens. A Living Machine reedbed system treats sewage on site, producing sufficiently clean water for us to put back in the rainwater tanks, recycling it for flushing toilets once again. When combined with water-efficient appliances such as dual flush low flush toilets, we are reducing mains water usage by 40%.

We have designed BedZED to make recycling easy. Each kitchen is fitted with containers for separating waste. There will be on-site composting of green waste, and we would have liked to keep a village pig or two were it not for the very tight regulations against feeding food waste to livestock. Ultimately we would like to see a system where residents are rewarded for reducing waste being sent to landfill, which is a cost currently borne by all through the local council community charge. If households generally are charged per kilogramme for waste they send to landfill, as is done in countries like Germany, those who recycle do not have to subsidise those who don't.

For BedZED to be really successful we have to demonstrate a lifestyle which is sustainable but also attractive. The spaces in homes and offices are full of natural light, well ventilated and stay at a constant comfortable temperature. People can avoid commuting, returning 3 hours per day to their lives—an extra 6 weeks a year, which they can spend as they wish. They have private gardens—something flat dwellers wouldn't have in an equivalent conventional development. They can save money by joining the car pool and by buying local organic food. They can feel part of a community and live a life which they know won't destroy the planet.

Maxine Chung was one of the first people to move into BedZED, Maxine commented: "I knew I wanted to live at BedZED within about five minutes of viewing the property. As well as the sustainable lifestyle that it promotes, the quality of the architecture and the attractive design are real plus points. It's great to see such a different and attractive development in south London. I'm already enjoying my new home at BedZED and I'm really looking forward to my future here. Living here should get even better when the rest of the residents move in. There will be a real sense of community here, which will enhance what is already a very attractive living environment."

Over the next 5 years, we will be monitoring the ecological footprints of residents at BedZED. Our initial work suggests that simply living at BedZED, without making any major changes to their lifestyle, allows residents to reduce their ecological footprint from the UK average of over 6 hectares, by about one third to around 4 hectares (see ecological footprint estimates in the colour centre pages). This saves us one of our three planets. Those residents who chose to take up some the additional green lifestyle services— like ZEDcars and local organic food—or are able to work in an energy and paper efficient office such as at BedZED, can reduce their footprint to the 1.9 hectare global fair share target. Thus they save the second planet and can live within the resources available on this one Earth.

Chapter 11

A Sustainable Future?

We hope that this Briefing not only defines the challenge ahead of us, but also demonstrates how developing within a bioregional framework can allow us to rise to that challenge. We will need to think consciously about how we can plan and implement more sustainable ways of living. To do this we will need to bring together all the necessary elements: eco-villages, local paper mills, sustainable fibre production, local organic food—and deliver them to customers as competitive sustainable products and services. We will need to create the economic and political climate where we can create a sustainable market, meeting more of our needs from local renewable and waste resources. We will need to foster equity between developing and developed countries and promote Fair Trade, not only because it is right, but also because it is in our own best interests. We have to learn to live as citizens sharing one small planet.

Over the past 8 years, we at BioRegional have been working to put into practice some of the solutions for sustainable living. None of this is rocket science (apart from solar panels, which were developed as part of the space programme!). There are resources and solutions all around us should we choose to adopt them.

As we look to the future, there are many more things that we at BioRegional want to do. Our projects are at different stages of development and implementation. We want to see our first MiniMill built and working and to developing the potential for BioRegional projects in China. We want to see our Local Paper recycling schemes strengthened and firmly established as the paper supply system of choice for the majority of companies. We would like to see most charcoal sold in the UK being UK-produced. In the longer term, we would like to expand on our work with groups in other countries who share our vision and commitment to delivering sustainable local products and services. In this way we can exchange knowledge and create international networks for local production.

For the 2002 World Summit we have twinned with the City of Johannesburg's EcoCity initiative. We are currently transferring some of the experiences in designing BedZED, and adapting the design for a township in South Africa called Ivory Park. Working with Bill Dunster Architects and Arup South Africa, and consulting with the local community, we have designed a community centre and terrace of demonstration homes which we hope will show how highly efficient buildings can be constructed cost-effectively even for those on low incomes.

Homes in South Africa, including those being built in townships, are being constructed at very low densities, causing urban sprawl and creating car dependence. This is a problem, especially in view of the lack of public transport systems. Township homes in particular are being built very rapidly as part of the African National Congress commitment to housing and they are better than the shacks they replace. However, in the rush to build them, little consideration is being given to how the city areas will function as communities. They are not being laid out to encourage 'village' centres, for shops and community facilities. Planning policy is separating areas into residential and commercial zones, furthering car dependence which is inevitably going to lead to increases in fossil fuel use.

Little consideration is given to energy efficiency in buildings. This means that in winter, low quality coal is being burned for heating, creating a smog problem, with 30% of children suffering respiratory problems because of it. Building at higher densities by constructing terraces of homes which share walls, means that money can be freed up to improve energy efficiency. We are experimenting with a recycling project, Iteke, in Ivory Park township to recycle waste polystyrene packaging into insulation for our demonstration homes. The insulation will keep buildings warm in winter and cool in summer. Insulation needs to be encouraged in commercial buildings as well, particularly to reduce air conditioning loads.

Building at higher densities will also free up land for communal areas, creating spaces for community interaction and nodes for public transport. There is great scope for creating a healthy relationship between the city and rural areas, for promoting urban agriculture and a sustainable infrastructure. We would advocate ensuring nutrients and water from sewage and green waste are used for growing crops locally. Consideration should also be given to a coherent recycling policy as a way to create local jobs. We have no doubt there is a lot that can be done simply using existing budgets but spending them differently. There are opportunities for Local Paper for Johannesburg and Local Paper for the Western Cape, for example. Local charcoal made using clean technology and wood from sustainably managed forests (or from clearance of invasive introduced tree species such as eucalyptus and wattle) could be a smokeless fuel for the townships. There is no shortage of possibilities, but it will take work and commitment to develop and implement the ideas.

We were very pleased that our designs for a community centre and demonstration homes were well received by the Ivory Park residents, in very great part due to the work of EcoCity over a number of years in explaining issues like the long-term benefits of building energy-efficient homes. We hope for a long term partnership with South Africa and to develop a full eco-village if our demonstration buildings are well received when they are constructed.

Back in the UK, we are working with partners such as Arup and the Peabody Trust to promote sustainable development in the Thames Gateway—the corridor east of London which will undergo major regeneration over the next few years. With sufficient vision and drive, a great opportunity exists to create a whole sustainable sub-region—based on zero waste and zero fossil energy principles—bringing together recycling industries, state-of-the-art building design, sustainable infrastructure and links to the farming hinterland. At this stage it remains a dream, but there are increasing numbers of Londoners who would like to see something of this sort happen.

In an evolution of our Local Paper for London scheme we will be piloting an innovative service with London Recycling: a pilot kerbside collection of waste paper from businesses, which also sells the recycled paper from the van. We have found that smaller offices and offices in central London do not always have storage space, nor can generate 10 sacks of paper for a free recycling collection, so they continue to dispose of their paper through trade waste sacks left on the pavement. A Westminster Council study found 40% of high grade paper in the total commercial waste. In this new service, offices will buy a paper recycling sack for half the price of a trade waste bag, which they put out on the street on a regular day and time. The pick-up van will have a separate compartment to deliver local recycled paper at the same time, cutting down on one more van journeys around London. In parts of London where bags cannot be left on the pavement for security reasons, we will develop a lobby storage and collection system.

We also have a proposal for a paper recycling mill which can economically re-process household graphics paper. This project has arisen from our study into the feasibility of meeting the government's Waste and Resources Action programme targets for graphics paper recycling.[1] Hundreds of thousands of tonnes of this material will be collected by local authorities as they implement recycling schemes in response to the EU Landfill Directive. Despite there being over 2 million tonnes available, household graphics paper is not currently an attractive raw material for the paper industry, as it comprises up to 50% clay, chalk and ink. When it is recycled, this 50% is discharged from the process as sludge, becoming a costly waste disposal problem. We have identified new technology, developed in the Netherlands, which cost effectively converts the sludge into cement and energy, and so could allow the economic recycling of this material.

We look forward to trialling the hemp textile processing technology in the UK and to the establishment of the new lavender farm. Working with Smart Moves, we are now expanding car clubs across seven London boroughs; and building on the expertise we have gained by building BedZED, we want to set up a London-wide reclaimed materials supply system. In another initiative, in order to bring costs down and improve efficiency, BioRegional and Bill Dunster Architects are developing pre-fabricated ZED components including wind cowls and walls with built-in service ducts.

Interest in BedZED has been enormous—and tours at our visitor centre are fully booked by a wide range of people from architects, planners and developers to interested members of the general public. It is starting to change perceptions, especially among industry professionals. But a gentle, informative approach is only part of the story. Conservatism in the property development and construction industries is still a barrier to progress. We are therefore working with planners in various local authorities to set planning briefs for sites which ask for developments with a BedZED level of environmental performance, so that any developer wishing to build on the site has to do it sustainably. Over time this will create a playing field where sustainability is a prerequisite, not an option. To be profitable, companies will have to be sustainable. In the UK, planners were disempowered when a more free market, laissez-faire approach was adopted through the 1980s. But the fact is that we can't build a sustainable or high quality future without strong planning direction. In our view, planners are going to be key—having the overview in terms of land-use, transportation and regional development. It is also our experience that planners will rise to the challenge when given the opportunity.

In another initiative we are supporting WWF-UK on a 1 million sustainable homes campaign. We hope to secure commitment from the government to get sustainable housing into the mainstream of the industry—creating the right environment through planning and financial instruments.

Bill Dunster Architects are taking the BedZED idea further. SkyZED is a concept for an eco-high rise, designed to generate more energy than it uses. Its four towers focus wind on central vertical axis wind turbines and its south-facing surfaces are clad in photovoltaic panels. Over time it will pay back the embodied energy of the materials employed in its construction, and will have a slightly beneficial effect on the earth's climate. Or, as Bill Dunster puts it, "It's a carbon-negative development." With Bill Dunster, BioRegional are investigating possible sites in London, Shanghai and Cape Town for the first SkyZED development.

Projects such as ours, and many others around the world, can help to accelerate the process of creating new, sustainable ways of living. Sometimes making the right environmental choice will save us money, such as many waste prevention measures. But the right environmental choice is not always the cheapest when we operate in a market which has not internalised all the environmental costs. While this remains the case, even the most well-meaning companies are limited all the time that their competitors are free to choose cheaper options. We will therefore need the courage to regulate the market sufficiently so that companies wishing to develop sustainable products and services are not held back by those who don't. It is not as though we don't already have social and environmental legislation. We can't employ child labour in the UK or use pesticides like endosulfin. But more needs to be done. In the UK we need to extend the scope of legislative standards—voluntary standards such as those of the Forest Stewardship Council for sustainable forestry should become legal standards. These standards will have to become international ones so that we are all competing on a level playing field. In this way we will avoid the situation where we say we won't use child labour in the UK, but buy goods made by children in other countries. If companies want to take advantage of cheaper labour in developing countries, let them employ adults and provide schools for the children as a legal requirement to trade. In the end we should have a world where consumers do still have choice, but only between sustainable alternatives.

We need better global governance in order to make lasting progress on global sustainability. Although it is difficult to achieve consensus between nations, we can only address certain problems such as CO_2 emissions at a global level. We have to be careful, however, in the approach we take to tackling the issues. When the UK government places a climate change energy levy on domestic industry, it might only succeed in giving an economic advantage to producers in another country who pay less for their energy. In Chapter 3 we gave the example of how a recycled paper mill in the UK is penalised in the face of competition from a virgin paper producer in Canada. Without a level playing field, progress is slow and unsure. It might be better to tax non-renewables at source.

The resource use insights given to us by concepts such as ecological footprinting can shed light on what is and what isn't sustainable. In a truly sustainable world, our ecological footprint budgets will be the same as our financial ones. That is the definition of a sustainable economy. When we reach this state, life will be a lot simpler.

Chapter 12

What You Can Do
to Live Bioregionally

If you would like to find out more about BioRegional projects and our partners please visit our website: www.bioregional.com. We also offer some suggestions below.

Individuals can:
- Plan long term to reduce your ecological footprint to the one planet level. When making key life decisions like deciding where to live, consider the effect on your ecological footprint.

- Think about the story behind any product or service you may purchase. Where has it come from and where will it go when you have finished with it? Will it contribute to planetary health? Avoid air freighted food, for instance.

- Choose locally produced, organic and Fair Trade products where possible.

- Think of your waste as a resource: re-use, recycle it or try and find someone who can use it.

- Think of yourself as a citizen of the planet, as well as of a nation.

- Think about co-operation rather than competition between nations and mutually respectful forms of development.

Businesses can:
- Think in terms of resource cycles and loops. Buy local recycled paper, recycle waste, promote energy efficiency, buy green electricity. In manufacturing consider using the principles of industrial ecology.

- Produce a sustainability action plan for the business and as part of every staff member's development plan to promote sustainability in their work.

- Think about the story behind any product or service your company sells or purchases.

- Work with local producers to develop sustainable local supplies.

- If you are a real enthusiast, create a sense of mission within the company, using sustainability, ecological footprinting and bioregional development as the cause.

Governments can:
- The National Audit Office can audit departmental accounts on the basis of sustainability and ecological footprinting, not simply money.

- Think about planetary interests at the same time as national interest.

- Work with industry and NGOs to create a sustainable market; ensure any regulatory or taxation reforms promote true environmental sustainability and global equity as a prerequisite.

- Work towards an economy where our financial budgets are equal to our ecological footprint budgets.

- Adopt bioregional development as a key way to deliver sustainability.

References

Chapter 1
1. WWF International (2000), 'Living Planet Report 2000', Avenue du Mont-Blanc, 1196 Gland, Switzerland.
2. Sale, K. (1991), *Dwellers in the Land: The Bioregional Vision*, New Society Publishers, Canada.

Chapter 2
1. WWF International (2000), 'Living Planet Report 2000', Avenue du Mont-Blanc, 1196 Gland, Switzerland.
2. The Royal Commission on Environmental Pollution (2000), 'Energy: The Changing Climate', Twenty-second report, The Stationery Office Limited, UK.
3. Meyer, A. (2000), *Contraction and Convergence*, Schumacher Briefing No 5.
4. Pretty, J. (1998), *The Living Land*, Earthscan Publications Ltd, UK.
5. Lomborg, B. (2001), *The Skeptical Environmentalist: measuring the real state of the world*, Cambridge University Press.
6. Report of Working Group 1 of the Intergovernmental Panel on Climate Change (2001), "Climate Change 2001: The Scientific Basis: Summary for Policy Makers'.
7. Mollison, B. (1990), *Permaculture: A Practical Guide for a Sustainable Future*, Island Press, Washington, DC.
8. Wackernagel, M. and Rees, W. (1996). *Our Ecological Footprint*, New Society Publishers, Canada.

Chapter 3
1. Porritt, J. (1997), speech at the 'Towards a sustainable paper cycle' seminar, International Institute for Environment and Development, London, UK.
2. *New Scientist* (November 1997), 'Burn me'.
3. Ecobilan Group (1998), 'Newsprint, a life cycle study', an independent assessment of the environmental benefits of recycling at Aylesford Newsprint compared with incineration.
4. Paper Federation (1998), 'An independent analysis of the policy options for the sustainable recovery of used newspapers in the UK'. Swansea, UK.
5. Hart, A. (2002), 'Bioregional Development: An analysis of the environmental implications of local "closed loop" graphics paper recycling'. Centre for Environmental Strategy, University of Surrey, Guildford.
6. Christopher Barr, CIFOR (2000), 'Profits on Paper: The Political-Economy of Fiber, Finance and Debt in Indonesia's Pulp and Paper Industries'.
7. Friends of the Earth (2002), 'Paper Tiger, Hidden Dragons 2: APRIL Fools', FoE UK www.foe.co.uk.
8. Greenpeace International (2001), 'Certified Destruction'.
9. Lomborg, B. (2001), *The Skeptical Environmentalist: measuring the real state of the world*, Cambridge University Press.

10. Food and Agriculture Organisation (2001), 'State of the World's Forests 2001' www.fao.org/forestry.

11. UNEP (2001), http//earthwatch.unep.net/forests/forestloss.html.

12. WorldWatch Institute (2000), *Vital Signs 2000-2001*, Earthscan.

13. www.paperloop.com (2002), Pulp and Paper International, Brussels.

14. McClaren, D., Bullock, S., Yousef, N. (1998), *Tomorrow's World, Britain's share in a sustainable future*, Friends of the Earth, London, Earthscan.

15. International Institute for Environment and Development (1996), 'Towards a Sustainable Paper Cycle', London.

16. Brown, L. (2001), *Eco Economy: building an economy for the Earth*, Earth Policy Institute US.

17. Riddlestone, S., Desai, P. Evans, M. Skyring, A. (1994), BioRegional Development Group, 'BioRegional Fibres, the potential for a sustainable regional paper and textile industry', UK.

18. Paper Federation of Great Britain (May 2001), 'Industry Facts 2000'.

19. Ibid.

20. Riddlestone et al (2002), BioRegional Development Group, 'A Graphic Waste of Paper—how we can meet UK targets for graphics paper recycling', London.

21. Forestry Commission (2001), 'Forestry Statistics 2001'.

22. *Environment Post* (2002), 'Good forest—bad forest', page 1, Issue 172, 9 May 2002.

23. Brown, L. (2001), *Eco Economy, building an economy for the Earth*, Earth Policy Institute USA.

24. As 20 above.

25. *Financial Times* (14 October 1999), 'Natural route to eco-friendly capitalism', reviewing *Natural Capitalism: the next Industrial Revolution* by Hawken, Lovins and Lovins, published by Earthscan 2000.

26. As 5 above.

27. 'Best Foot Forward' (2002), ecological footprint of local recycled paper, Oxford, (unpublished).

Chapter 4

1. Schumacher, E.F. (1973), *Small is Beautiful: economics as if people mattered*.

2. IIED, (1996), *Towards a Sustainable Paper Cycle*, London.

3. Chinese Ministry of Foreign Trade and Economic Affairs MOFTEC (2000).

4. Professor Li (2002), Chinese Ministry of Light Industry, personal communication.

5. Friends of the Earth (2002), 'Paper Tiger, Hidden Dragons 2: APRIL Fools', www.foe.co.uk.

6. Pulp and Paper International (2000), 'International fact and price book', Brussels.

Chapter 5

1. Rackham, O. (1986), *The History of the Countryside*, J.M. Dent and Sons Ltd, London.

2. Budd, P. (1993), 'Recasting the Bronze Age', *New Scientist* No. 1896, pp33-37.

3. Hart, A. (1997), 'Charcoal for Barbecue Consumption within the UK: A Transport Energy Analysis'. BSc dissertation, Queen Mary and Westfield College, London.

4. UNCTAD (2000), 'Least Developed Countries Report', UNCTAD, New York and Geneva.

5. UNEP (2001), reported in *The Environment Post*, Issue 162, Pathway UK Ltd, Basildon, Essex, UK.

6. Sustain—the alliance for better food and farming (2001), 'Eating Oil: Food Supply in a Changing Climate', Sustain / Elm Farm Research Centre Publication.

7. South African Airways (2002), In-flight magazine May 2002.

Chapter 6
1. Simmons, Craig, 'Best Foot Forward' personal communication.
2. Sustain—the alliance for better food and farming (2001), 'Eating Oil—Food Supply in a Changing Climate', Sustain / Elm Farm Research Centre Publication.
3. Pretty, Professor J. (1998), *The Living Land*, Earthscan Publications Ltd, UK.
4. Davies, N., Desai, P. and Hughes, D. (2000), 'Local Sourcing Initiative', BioRegional Development Group.
5. Policy Commission on the future of Farming and Food (2002), 'Farming and food a sustainable future'. www.cabinet-office.gov.uk/farming
6. 'Energy use in organic farming systems', ADAS Consulting for MAFF, Project OFO 182, DEFRA, London 2001
7. Quoted in: Sustain—the alliance for better food and farming (2001), 'Eating Oil—Food Supply in a Changing Climate', Sustain / Elm Farm Research Centre Publication
8. Lucas, Dr. C. (2001), 'Stopping the great food swap—relocalising Europe's food supply'. The Greens/European Free Alliance, The European Parliament, Brussels.

Chapter 7
1. *The Australian*, June 8 1992.
2. Erin Gill, 'Not just Patagucci man, *Green Futures*, UK, July/August 2000.
3. Barbara Dinham, Pesticides Trust, personal communication.
4. WWF International (1999), 'The impact of cotton on freshwater resources and ecosystems', Switzerland.
5. Textile statistics bureau figures UK, cotton imports 1992.
6. *The Ecologist* Vol. 22 No. 4, Jul/Aug 1992.
7. The Aral sea pages, www.dfd.dlr.de/app/land/aralsee (sic).
8. Perrera, J. (1993), 'A sea turns to dust', *New Scientist* 23 October 1993, pp24-27.
9. *New Scientist* 124:1691:22 18 November 1989. Micklin P.P. 'Desiccation of the Aral Sea, water management disaster in the Soviet Union'.
10. Riddlestone, S. (1992), 'The Cotton Story', BioRegional Development Group, UK.
11. Scarborough et al (1989), 'Acute effects of community exposure to cotton defoliants', Archives of environmental health 44(6) 355-360, Pimental D. and Levitan L. 1986.
12. Gilham F. et al (1995), 'Cotton production prospects for the next decade', World Bank technical paper number 287. World Bank, Washington DC.
13. *Xinhua* (2001), 'China's biggest cotton zone hit by pests', 13 September 2001.
14. Abrams, F., Astill, J. (2001), 'Story of the Blues' *The Guardian*, London, 29 May 2001.
15. Johnston, J. 'Sold for £10, heartbreaking story behind Africa's child slave trade', *The Mirror*, London, April 21 2001.
16. George, S. (1988), *A fate worse than debt*, Penguin.
17. US Department of Agriculture (2002), www.fas.usda.gov/cotton/circular/2002/03/master2.htm.
18. Clare Short (2001), Secretary of State for International Development speaking at the Ministerial round table on trade and the least developed countries, London 19 March 2001.
19. Oxfam (2002), 'Loaded against the poor', www.oxfam.org.uk.
20. *The Australian*, June 8 1992.
21. Riddlestone, S., Desai, P., Evans, M., Skyring, A. (1994), BioRegional Development Group, 'BioRegional Fibres, the potential for a sustainable regional paper and textile industry', UK.
22. Alden et al, (1996), 'Industrial hemp's double dividend: a study for the USA', University of Melbourne, Dept of Economics, Australia, July 1996.
23. Riddlestone S., Franck, R. , Wright, J., (1995), 'Hemp for Textiles: growing our own clothes', BioRegional Development Group, UK.

Chapter 8
1. Mayor's Draft Spatial Development Strategy (2001), Greater London Authority.
2. Barlow, J. et al (2002), 'Land for Housing', published for the Joseph Rowntree Foundation by York Publishing Services, York.
3. BBC News Online, 28 January 2000, 'Clothing workers given notice'.
4. BBC News Online, 10 January 2002, UK 'faces tropical disease threat'.
5. London Tree Officers Association (2000), 'London Bioenergy Report'.

Chapter 9
1. Goldsmith, E. and Mander, J. (2001), *The Case against the Global Economy*, Earthscan Publications Limited, UK.
2. Monbiot, G. (2001), *Captive State: the corporate takeover of Britain*, Macmillan.
3. Roddick, A. (2001), *Take it personally*, www.anitaroddick.com.
4. Soros, G. (2002), *George Soros on Globalisation*, PublicAffairs LLC.
5. *Green Futures* (Jan/Feb 2002), interview with Adair Turner.
6. Maunder, P. et al. (1995), *Economics Explained*, Third Edition, Collins Educational, London.
7. Soros, G. *Crisis in global capitalism*.
8. Stott, R. (2000), *The Ecology of Health*, Schumacher Briefing No 3, Green Books, Totnes, Devon, UK.

Chapter 10
1. Girardet, H. (1999), *Creating Sustainable Cities*, Schumacher Briefings No 2, Green Books, Totnes.
2. Ibid.
3. Rogers, R. (1997), *Cities for a Small Planet*, Faber and Faber Ltd, London.

Chapter 11
1. Riddlestone et al (2002), BioRegional Development Group, 'A Graphic Waste of Paper—how we can meet UK targets for graphics paper recycling'. London.

SCHUMACHER BRIEFINGS

The Schumacher Briefings are carefully researched, clearly written booklets on key aspects of sustainable development, published approximately three times a year. They offer readers:

• background information and an overview of the issue concerned
• an understanding of the state of play in the UK and elsewhere
• best practice examples of relevance for the issue under discussion
• an overview of policy implications and implementation.

The first Briefings are as follows:

No 1: Transforming Economic Life: A Millennial Challenge by James Robertson

Chapters include Transforming the System; A Common Pattern; Sharing the Value of Common Resources; Money and Finance; and The Global Economy. Published with the New Economics Foundation.

No 2: Creating Sustainable Cities by Herbert Girardet

Shows how cities can dramatically reduce their consumption of resources and energy, and at the same time greatly improve the quality of life of their citizens. Chapters include Urban Sustainability, Cities and their Ecological Footprint, The Metabolism of Cities, Prospects for Urban Farming, Smart Cities and Urban Best Practice.

No 3: The Ecology of Health by Robin Stott

Concerned with how environmental conditions affect the state of our health; how through new processes of participation we can regain control of what affects our health, and the kinds of policies that are needed to ensure good health for ourselves and our families.

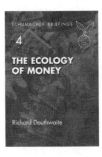

No 4: The Ecology of Money
by Richard Douthwaite

Explains why money has different effects according to its origins and purposes. Was it created to make profits for a commercial bank, or issued by government as a form of taxation? Or was it created by users themselves purely to facilitate their trade? This Briefing shows that it will be impossible to build a just and sustainable world until money creation is democratized.

No 5: Contraction & Convergence: The Global
Solution to Climate Change by Aubrey Meyer

The C&C framework, which has been pioneered and advocated by the Global Commons Institute at the United Nations over the past decade, is based on the thesis of 'Equity and Survival'. It seeks to ensure future prosperity and choice by applying the global rationale of precaution, equity and efficiency in that order.

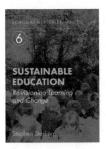

No 6: Sustainable Education: Revisioning
Learning & Change by Stephen Sterling

Education is largely behind—rather than ahead of—other fields in developing new thinking and practice in response to the challenge of sustainability. The fundamental tasks are to
• critique the prevailing educational and learning paradigm, which has become increasingly mechanistic and managerial
• develop an ecologically informed education paradigm based on humanistic and sustainability values, systems thinking and the implications of complexity theory. An outline is given of a transformed education that can lead to transformative learning.

No 7: The Roots of Health
by Romy Fraser and Sandra Hill

The advancements of modern medicine provide a sophisticated but mechanistic approach to health. Dazzled by its progress, we have lost touch with the simple remedies and body wisdom that were once a part of every household. By understanding the roots of health, we can begin to reclaim this wisdom, and to heal ourselves, our society and our environment.

All the above Briefings are 80/96pp, and cost £5 each plus postage.

THE SCHUMACHER SOCIETY
Promoting Human-Scale Sustainable Development

The Society was founded in 1978 after the death of economist and philosopher E. F. Schumacher, author of seminal books such as *Small is Beautiful*, *Good Work* and *A Guide for the Perplexed*. He sought to explain that the gigantism of modern economic and technological systems diminishes the well-being of individuals and communities, and the health of nature. His works has significantly influenced the thinking of our time.

The aims of the Schumacher Society are to:

• help assure that ecological issues are approached, and solutions devised, as if people matter, emphasising appropriate scale in human affairs;

• emphasise that humanity can't do things in isolation. Long-term thinking and action, and connectedness to other life forms, are crucial;

• stress holistic values, and the importance of a profound understanding of the subtle human qualities that transcend our material existence.

At the heart of the Society's work are the Schumacher Lectures, held in Bristol every year since 1978, and now also in Manchester, Edinburgh and elsewhere in the UK. Our distinguished speakers, from all over the world, have included Amory Lovins, Herman Daly, Petra Kelly, Jonathon Porritt, James Lovelock, Wangari Maathai, Matthew Fox, Ivan Illich, Fritjof Capra, Arne Naess, Maneka Gandhi, James Robertson and Vandana Shiva.

Tangible expressions of our efforts over the last 20 years are: the Schumacher Lectures; Resurgence Magazine; Green Books publishing house; Schumacher College at Dartington, and the Small School at Hartland, Devon. The Society, a non-profit making company, is based in Bristol and London. We receive charitable donations through the Environmental Research Association in Hartland, Devon.

Schumacher Society Members receive:

• a free lecture ticket for either Bristol, Manchester or Edinburgh
• the Schumacher Newsletter
• the catalogue of the Schumacher Book Service
• information about Schumacher Briefings
• information about Schumacher College Courses

**The Schumacher Society, The CREATE Environment Centre,
Smeaton Road, Bristol BS1 6XN Tel/Fax: 0117 903 1081
schumacher@gn.apc.org www.schumacher.org.uk**